NCIDQ PROFESSIONAL

PRACTICE QUESTIONS AND MOCK EXAM

THIRD EDITION

DAVID KENT BALLAST, FAIA
NCIDQ CERTIFICATE NO. 9425

Report Errors for This Book

PPI is grateful to every reader who notifies us of a possible error. Your feedback allows us to improve the quality and accuracy of our products. Report errata at **ppi2pass.com/errata**.

NCIDQ Professional Practice Questions and Mock Exam
Third Edition

Current release of this edition: 1

Release History

date	edition number	revision number	description
Jul 2021	3	1	New edition.

Printed in the United States of America.

PPI
ppi2pass.com

ISBN: 978-1-59126-844-4

TABLE OF CONTENTS

PREFACE AND ACKNOWLEDGMENTS ..... **v**

INTRODUCTION ..... **vii**

How to Use This Book vii

About the NCIDQ Exam viii

How to Register for the Exam xiii

IDPX Tips xv

What to Do After the Exam xvi

HOW SI UNITS ARE USED IN THIS BOOK ..... **xvii**

CODES, STANDARDS, AND REFERENCES FOR THE EXAM ..... **xix**

PRACTICE QUESTIONS **1**

1. Project Assessment and Sustainability 1
2. Project Process, Roles, and Coordination 7
3. Professional Business Practices 13
4. Code Requirements, Laws, Standards, and Regulations 17
5. Integration with Building Systems and Construction 26
6. Integration of Furniture, Fixtures, and Equipment 32
7. Contract Administration 36

MOCK EXAM 43

1. Project Assessment and Sustainability 45
2. Project Process, Roles, and Coordination 50
3. Professional Business Practices 56
4. Code Requirements, Laws, Standards, and Regulations 59
5. Integration with Building Systems and Construction 66
6. Integration of Furniture, Fixtures, and Equipment 71
7. Contract Administration 75

MOCK EXAM ANSWERS 81

1. Project Assessment and Sustainability 85
2. Project Process, Roles, and Coordination 92
3. Professional Business Practices 97
4. Code Requirements, Laws, Standards, and Regulations 99
5. Integration with Building Systems and Construction 107
6. Integration of Furniture, Fixtures, and Equipment 113
7. Contract Administration 116

PREFACE AND ACKNOWLEDGMENTS

Every five years, the Council for Interior Design Qualification (CIDQ)[1] commissions a practice analysis to ensure the exam continues to accurately assess the knowledge and skills interior designers need to practice responsibly and to protect the health, safety, and welfare of the public. Results of the practice analysis are reported in the council's publication *Analysis of the Interior Design Profession*. The council administers three exams (the Interior Design Fundamentals Exam (IDFX), the Interior Design Professional Exam (IDPX), and the Interior Design Practicum) that cover the content areas identified in the analysis as being the essential tasks, knowledge, and skills of interior designers in the United States and Canada. I wrote *NCIDQ Professional Practice Questions and Mock Exam* to be consistent with IDPX's content areas so that you will use it to prepare for and pass the IDPX.

In this third edition, I have reorganized questions and added new questions to be consistent with the 2021 NCIDQ exam content areas for IDPX. Also included are new item types that the NCIDQ exam may use, including check-all-that-apply, fill-in-the-blank, hot spot, and drag-and-place. You will also find more questions based on illustrations.

I would like to thank Holly Williams Leppo, AIA, NCIDQ Certificate No. 19182, who reviewed the problems for validity and offered many good suggestions for additions and improvements. Many people have helped in the production of this book. I would like to thank Megan Synnestvedt, product director; Nicole Evans, senior product manager; Meghan Finley, content specialist; Beth Christmas, project manager; Indira Prabhu Kumar, editor; Michael Wordelman, editor; Scott Marley, editorial manager; Sean Woznicki, production editor; Richard Iriye, typesetter; Tom Bergstrom, production specialist; Kim Burton-Weisman, proofreader; Nikki Capra-McCaffrey, production manager; Sam Webster, publishing systems manager; and Grace Wong, director of editorial operations.

While I had a lot of help with this book, any mistakes are mine alone. If you find a mistake, please submit it through PPI's errata page at **ppi2pass.com/errata**. The error will be revised accordingly.

David Kent Ballast, FAIA
NCIDQ Certificate No. 9425

[1]In this Preface, I refer to the Council for Interior Design Qualification as "the council," and the National Council for Interior Design Qualification exam as "the exam."

INTRODUCTION

HOW TO USE THIS BOOK

NCIDQ Professional Practice Questions and Mock Exam was developed in accordance with the Council for Interior Design Qualification (CIDQ[1]) Interior Design Professional Exam (IDPX) content areas. Though the exact questions in this book won't be found on the actual exam, they are like IDPX questions in terms of their format, level of difficulty, and the content areas they cover.

This book is dual-purpose and has two parts: 100 realistic practice questions and a 175-question mock exam. Both the practice questions and the mock exam are divided by content area to make it easy to focus your studying on content areas you are less familiar with.

Use the practice questions in the first part of this book to become familiar with the exam's content areas. Answers to the practice questions are given directly after the questions so you can immediately check your comprehension of a subject. Use the *NCIDQ Interior Design Reference Manual* (IDRM, also published by PPI) to research less familiar topics or questions answered incorrectly. Taking the time to research less familiar topics will help to strengthen your understanding and ensure exam readiness.

Once you are comfortable with IDPX's content areas, use the mock exam in the second part of this book to simulate the exam experience. Put away study materials and references, set a timer for four hours, and answer as many of the mock exam questions as you can within the time limit. Complete the entire mock exam, trying to answer each question in no more than about 1 minute 15 seconds. If a question seems to be taking longer, skip it. Note unanswered questions to go back to later and answer. This should leave a reserve of at least 20 minutes at the end of the exam session to review and at least guess at the questions left unanswered. Reread the questions you were able to answer only if there is extra time after guessing at all questions you marked to revisit. Be sure to mark an answer for every

[1]The Council for Interior Design Qualification is referred to as "the council" throughout this Introduction, while the National Council for Interior Design Qualification exam is referred to as "the exam."

question. On IDPX, unanswered questions are counted as wrong, so when you are unsure of an answer, make an educated guess among the most likely options.

If time runs out before you are able to complete all mock exam questions, make a note of the last question you worked on within the time limit, but continue to complete the entire practice exam. Keep track of time to determine how much faster you would need to work to finish the actual exam within four hours.

After completing the mock exam, check your selected answer options against the answer key. Each correct question is worth one point. There is no penalty for questions answered incorrectly. To score the mock exam, multiply the number of correct questions by 4.6.[2]

Use the fully explained answers at the end of the mock exam as a learning tool. Take note of the content areas with the most questions answered incorrectly, and focus the majority of your exam review around those.

ABOUT THE NCIDQ EXAM

The exam is divided into three sections: the Interior Design Fundamentals Exam (IDFX), the Interior Design Professional Exam (IDPX), and the Interior Design Practicum (PRAC). All three exams are administered via computer. IDFX contains questions that test the knowledge gained in school (e.g., programming and site analysis; design application and human behavior; design communication; life safety and universal design; interior building materials and finishes; technical specifications for furniture, fixtures, and equipment and lighting; construction document standards; and professional development and ethics). IDPX contains questions that test knowledge gained through work experience (e.g., project assessment and sustainability; project process, roles, and coordination; professional business practices; code requirements, laws, standards, and regulations; building systems and construction; integration of furniture, fixtures, and equipment; and contract administration). The Practicum contains exercises that test practical interior design knowledge (e.g., programming, planning, and analysis; code requirements, laws, standards, and regulations; integration with building systems; and contract documents). There is some content area overlap between the three exams. See "Eligibility" later in this Introduction for information on the requirements for each exam. All three exams are administered in April and October. Through Prometric, you can schedule your exams for any dates within those two months.

IDPX is four hours long with 175 questions (150 of which are scored). The remaining 25 questions are used for developmental purposes and are not scored. (These questions are not identified in advance.) Both IDFX and IDPX are scored on a scale of 200 to 800 points, with 500 being the minimum number of points needed to pass. Points are not deducted for questions answered incorrectly.

Exam Problem Types

There are several types of problems on IDFX, IDPX, and in this book.

- multiple-choice problems

[2]On the actual exam, the council includes 25 unscored, experimental questions. To score the mock exam to parallel a real exam score, randomly choose 25 questions to leave unscored. Among the remaining 150 questions, calculate the number of correct questions and multiply this number by 5.3.

- check-all-that-apply problems
- fill-in-the-blank problems
- hot spot problems
- drag-and-place problems

Multiple-Choice Problems

Multiple-choice problems have two types. One type of multiple-choice problem is based on written, graphic, or photographic information. You will need to examine the information and select the correct answer from four given options. Some problems may require calculations. A second type of multiple-choice problem describes a situation that could be encountered in actual practice. Drawings, diagrams, photographs, forms, tables, or other data may also be given. The problem requires you to select the best answer from four options.

Multiple-choice problems often require you to do more than just select an answer based on memory. At times it will be necessary to combine several facts, analyze data, perform a calculation, or review a drawing.

Check-All-That-Apply Problems

Check-all-that-apply problems are a variation of a multiple-choice problem, where six options are given, and you must choose all the correct options. The problem tells how many of the options are correct, from two to four. You must choose all the correct options to receive credit; partial credit is not given.

Fill-in-the-Blank Problems

Fill-in-the-blank problems require you to fill in a blank with a value that you have derived from a table or a calculation.

Hot Spot Problems

Hot spot problems are used to assess visual judgment, evaluation, or prediction. Hot spot problems include the information needed to make a determination, along with an image (e.g., diagram, floor plan) and instructions on how to interact with the image. The problems will indicate that you should place a single target, also known as a hot spot icon, on the base image in the correct location or general area. On the exam, you will place the target on the image by moving the computer cursor to the correct location on the image and clicking on it. You will see crosshairs to help you position the point of click. You will be able to click on an alternate spot if you think your first choice is not correct. Your choice is not registered until you exit the problem. You can click anywhere within an acceptable area range and still be scored as correct.

Drag-and-Place Problems

Drag-and-place problems are similar to hot spot problems, but whereas hot spot problems involve placing just one target on the base image, drag-and-place problems involve placing two to six design elements or text onto the base image. Drag-and-place problems are used to assess visual judgment or evaluation with multiple pieces of information. The problem statement describes what information is to be used to make the determination, and provides instructions on how to interact with the image or graphic item.

A drag-and-place problem, for example, may require you to drag and place design elements such as wallboard trim onto the base image. On the exam, you will use the computer cursor to place the elements on the image by clicking and holding elements and dragging and releasing the elements on the correct location on the image. Depending on the question, you may use an element more than once or not at all. This type of question also provides an acceptable area range for placing the elements. The range may be small for questions about a detail or large for something like a floor plan.

Both IDFX and IDPX are machine-graded.

For more information and tips on how to prepare for IDFX, consult the *Interior Design Reference Manual* or visit PPI's website, **ppi2pass.com** (keywords: NCIDQ, IDRM).

The Practicum includes individual exercises that will require you to interpret a program, translate it into schematics, produce plan drawings, and develop appropriate specifications and schedules. Each exercise is scored by two NCIDQ graders. Graders give each exercise a score of 0, 1, 2, 4, or 5. There is no score of 3. Scores of 0, 1, and 2 are failing, and 4 and 5 are passing. If the exercise receives one passing and one failing score, a third NCIDQ grader will review the exercise. The two scores for each exercise are added together and multiplied by a weighting factor. The resulting value becomes a percentage of the raw score.

IDFX is three hours long with 125 questions (100 of which are scored), and IDPX is four hours long with 175 questions (150 of which are scored). The remaining 25 questions in each are used for developmental purposes and are not scored. (These questions are not identified in advance.) The Practicum is four hours long and includes 120 questions divided across three case studies.

IDPX Content Areas

The council uses its *Analysis of the Interior Design Profession* to develop the exam content areas, which cover the knowledge and skills that interior designers must possess to protect public health, safety, and welfare. For IDFX and IDPX, the number of questions in each content area is related to that content area's relative importance, as evaluated through survey responses from practicing interior designers.

The content areas for IDPX are as follows. The percentage of weight for each content area is given in parentheses.

Interior Design Professional Exam (IDPX)

I. Project Assessment and Sustainability (15%)

Ability to understand and analyze

- square footage standards (e.g., building codes, BOMA calculations and terminology)

Demonstrate understanding of

- environmental and wellness attributes (e.g., energy and water, conservation, renewable resources, indoor air quality, resiliency, active design)
- existing conditions analysis (e.g., hazardous materials, seismic, accessibility, construction type, occupancy type)
- project drivers (e.g., stakeholder requirements, space usage, preferred culture and branding, goals and objectives, budget)

II. Project Process, Roles, and Coordination (15%)

Understand and identify

- scope of project team members (e.g., architects, engineers, specialty consultants, contractors, construction managers)
- role of stakeholders (e.g., management, identification, level of interest, level of influence, managing expectations)

Demonstrate understanding of

- project budgeting/tracking (e.g., value engineering, alternates, timing and responsibility)
- critical path (e.g., design milestones, sequencing, design phases, deliverables)
- design phase criteria (e.g., deliverables, approval, sign-off, quality control, meeting project goals and objectives)
- allied professionals' drawings (e.g., mechanical, electrical, and structural engineering, architecture, security, specialty consultants)
- specification types and format (e.g., prescriptive, performance, proprietary, divisions)
- phased construction plan
- post-occupancy evaluation (e.g., metrics, timing, scope, analyzing data, evaluating criteria, commissioning, employee surveys)

III. Professional Business Practices (10%)

Demonstrate understanding of

- scope of practice (e.g., legal liability, laws and regulations, certification vs. licensure, practice and title act)
- business structures (e.g., LLC, joint ventures, sole proprietor, partnership, corporation)
- business management (e.g., applicable taxes, accounting, liability and insurance)

Ability to understand and develop

- proposals (e.g., time and fee estimation, request for proposals, process, project scope, presentation, exclusions, and add services)
- contracts (e.g., legal considerations, liabilities, terms and conditions)
- project budgeting principles and practices

IV. Code Requirements, Laws, Standards, and Regulations (20%)

Demonstrate understanding of

- environmental regulations (e.g., indoor air quality, energy conservation, material conservation, water conservation)
- reference standards and guidelines (e.g., ADA/Accessibility, BIFMA, ASHRAE, OSHA, NFPA, IBC)
- zoning and building use
- permit requirements (e.g., processes, timing, awareness of jurisdictional differences)

V. Integration with Building Systems and Construction (15%)

Demonstrate understanding and application of

- structural systems (e.g., load bearing, non-load bearing, steel, concrete, post-tension)
- plumbing systems (e.g., low flow, waterless, filtration, water metering, gray water)
- fire protection systems (e.g., sprinklers, strobes, alarms, extinguishers, smoke and heat detectors)
- low voltage systems (e.g., data and communication, security, A/V)
- mechanical systems (e.g., types of systems, coordination with ceiling plans, indoor air quality)
- monitoring systems (e.g., building automation systems)
- installation methods (e.g., sequencing of work)
- building construction types (e.g., wood, steel, concrete)
- building components (e.g., doors, windows, wall assemblies, hardware, glazing assemblies)
- vertical and horizontal systems of transport (e.g., stairs, elevators)
- lighting systems (e.g., fixtures, zoning, sensors, daylighting, circadian rhythms, calculations, distribution, energy efficiency)
- electrical systems (e.g., outlet placement, switching, GFI, occupancy sensors)
- acoustical systems (e.g., sound masking, NRC, STC, CAC, AC, sound batting, wall types and ceiling elements)

VI. Integration of Furniture, Fixtures, and Equipment (10%)

Identify and apply appropriate

- product components (e.g., system furniture vs. ancillary furniture, power integration of furniture, acoustic panels vs. non-acoustic panels, modular wall systems)

Demonstrate understanding of

- equipment integration (e.g., appliances or specialty equipment within the design, accessibility and code compliance)
- parameters of maintenance (e.g., warranties, manuals, cleaning protocols, documents)
- processes for procurement, delivery, and installation (e.g., sequencing, purchase orders, prepayment requirements, customer's own material, liabilities, shop drawings, lead time)

Ability to conduct and communicate

- budgeting and cost estimating (e.g., quantity takeoffs, product cost, install cost, overage, attic stock, life cycle costing, return on investment)

VII. Contract Administration (15%)

Demonstrate understanding of

- application of documentation and procedures (e.g., transmittals, contemplative change orders, change directive, change order, addenda, bulletin, purchase orders, request for information (RFIs))

- project accounting (e.g., payment schedules, invoices, contractor pay applications and approvals)

Ability to lead

- project meetings (e.g., management, protocol, minutes)

Demonstrate understanding and utilization of

- site visits and field reports
- shop drawings and submittals
- construction mock-ups
- punch lists/deficiency lists

HOW TO REGISTER FOR THE EXAM

Registering for the exam is a multi-step process. First, you must meet the exam's eligibility requirements, then submit an application and have it accepted. Confirm your eligibility, and begin the application process well before the exam date.

Eligibility

You may take IDFX after you have met the minimum education requirements. You may meet the education requirements by receiving a bachelor of arts or master of fine arts degree from an interior design program accredited by the Council for Interior Design Accreditation (CIDA); a bachelor of arts degree or higher from an interior design program not accredited by CIDA with at least 60 semester (or 90 quarter) credits in interior design coursework; a bachelor of arts degree or higher in another major with at least 60 semester (or 90 quarter) credits of interior design coursework that led to a diploma, degree, or certificate; an associate of arts degree with at least 40 semester (or 60 quarter) units in interior design; or a bachelor of arts or master of fine arts degree from an architecture program accredited by the National Architectural Accrediting Board (NAAB) or Canadian Architectural Certification Board (CACB).

You may take IDPX and the Practicum after you have met the work experience requirements given in this section. Work experience for which academic credit was received will not count toward the hours required. Passing IDFX is not a prerequisite to taking IDPX and the Practicum, although you must receive a passing score on all three exam sections to earn the NCIDQ certificate.

To be eligible for IDPX and the Practicum with a bachelor of arts or master of fine arts degree from an interior design program accredited by CIDA, you must complete 3520 hours of work experience.[3] At least 1760 of those hours must be accrued after you have met your education requirements.

[3]Hours worked under an NCIDQ certificate holder, a licensed/registered interior designer, or an architect who offers interior design services count as "qualified" work experience, and accrue at 100%. Hours worked in alternative situations accrue at lower rates. Hours worked under direct supervision by an interior designer who is not registered, licensed, or an NCIDQ certificate holder accrue at a rate of 75%; work sponsored, but not directly supervised, by the same individual accrues at a rate of 25%. Work hours not supervised by a designer (i.e., supervised by someone other than a designer, or not supervised at all in the case of self-employment) accrue at a rate of 25%. The Interior Design Experience Program (IDEP) offered by the council satisfies the work experience requirement.

To be eligible for IDPX and the Practicum with a bachelor of arts degree or higher from an interior design program that is *not* accredited by CIDA, you must complete 3520 hours of work experience. At least 1760 of those hours must be accrued after you have met your education requirements. In addition, at least 60 semester (or 90 quarter) credits must be in interior design coursework.

To be eligible for IDPX and the Practicum with a bachelor of arts degree or higher in another major, you must complete 3520 hours of work experience. At least 1760 of those hours must be accrued after you have met your education requirements. In addition, at least 60 semester (or 90 quarter) credits of interior design coursework must be completed and must have resulted in a diploma, degree, or certificate.

To be eligible for IDPX and the Practicum with an associate of arts degree of 60 semester (or 90 quarter) units in interior design, you must complete 5280 hours of work experience. All these hours must be accrued after you have met your education requirements.

To be eligible for IDPX and the Practicum with an associate of arts degree of 40 semester (or 60 quarter) units in interior design, you must complete 7040 hours of work experience. All these hours must be accrued after you have met your education requirements.

To be eligible for IDPX and the Practicum with a bachelor of arts or master of fine arts degree from an architecture program accredited by NAAB or CACB, you must complete 5280 hours of work experience. All these hours must be accrued after you have met your education requirements.

Applying

Exam dates and application deadlines are listed on PPI's website at **ppi2pass.com** (keyword: NCIDQ). The first step in the application process is submitting the online application form and application fee through MyNCIDQ, a section of CIDQ's website (cidq.org). For IDFX, gather the following supporting materials and apply online through MyNCIDQ. Unopened official transcripts should be mailed.

- *Official transcripts:* Download the transcript request form from the council's website and submit it to the registrar (along with any fees the registrar requires) for each college or university you attended. The registrar will return the official transcript to you in a sealed envelope which must be included, *unopened*, in your package of supporting materials.

For IDPX and the Practicum, the following items are also needed.

- *Work experience verification forms:* Submit a separate form for each position you held. Complete the direct supervision work experience verification form for work experience you completed under a direct supervisor. For work experience not directly supervised by a design professional (for example, in the case of self-employment), complete the sponsored work experience verification form. A sponsor is a design professional who can verify work experience, but was not a direct supervisor and may not have had direct control over or detailed knowledge of your work.

Do not include any additional materials in your package of supporting materials. Any additional documents you include will be discarded. Be sure to have all materials submitted by the relevant deadline; the council will not review a partial application or an application once the deadline has passed.

Registering

If your application is accepted by the council, you will be notified through MyNCIDQ. If your application is not accepted, you may need further education and work experience to fulfill the requirements. If this process takes more than one year, you will need to resubmit your entire application.

If your application is accepted, you may register for as many exam sections as you were accepted for. If you choose not to register for this exam cycle, you will remain an active candidate and will receive email notifications about registering for the next exam cycle. Fees are listed in the registration guide (a brochure available for download from the council's website) for each exam section.

Registering Through Prometric

To register for IDFX or IDPX, go to Prometric's website (prometric.com/NCIDQ) or call the toll-free number. Choose the exam location and date from those available. All locations accept registrations up to a day before the exam; some locations accept same-day registration. Registration fees are payable by credit card only.

Prometric will send you an email with your registration confirmation and test center information. Bring two forms of identification with you on examination day. One of these must be a government-issued photo ID.

Registering Through MyNCIDQ

To register for the Practicum, go to MyNCIDQ and click on "Exam Registration." Complete the confidentiality agreement and statement of responsibility. Registration fees are payable online by credit card or through the mail by check (this must be received in time for processing).

The council will send you an email with further exam information after receipt of payment and an email with your letter of admission at least two weeks prior to the exam. Print the letter of admission and present it with a government-issued photo ID on examination day.

IDPX TIPS

Consider the following tips while taking IDPX.

- Try to complete each question in no more than 1 minute 15 seconds to leave a reserve of about 20 minutes to guess at unanswered questions at the end of the exam session.
- Eliminate any obviously incorrect options before attempting to guess. The chances are better between two choices than among four.
- Look for an exception to a rule or a special circumstance that makes the obvious, easy response incorrect. Although there may be a few easy and obvious questions, it's more likely that a simple question has a level of complexity that is not immediately obvious.
- Take note of absolute words such as "always," "never," or "completely." These words often indicate some minor exception that can turn what reads like a true statement into a false statement, or vice versa.
- Watch for words like "seldom," "usually," "best," or "most reasonable." These words generally indicate that some judgment will be involved in answering the question, so look for two or more options that may be very similar.

- If a question appears to be fundamentally flawed, make the best choice possible under the circumstances. Flawed questions do not appear often on the exam, but when they do, they are usually discovered by the council in the grading process. These questions will not negatively impact your score.

WHAT TO DO AFTER THE EXAM

Score notifications for IDFX and IDPX will be sent within eight weeks after your test date. Score notifications for the Practicum will be sent within 14 weeks after your test date. If you pass the Practicum, you will receive a score report that indicates "Pass." If you fail the Practicum, you will receive a score report that indicates "Fail," along with a list of the exercises receiving failing scores. If you fail one or two of the three exam sections, you will only need to retake the failed section(s). Your state or province may impose restrictions on the number of years you can wait between taking different sections of the exam.

The council will issue you a certificate and a certificate number when you pass all three exam sections. The certificate number will be unique, and is used to distinguish designers within the interior design field. To identify yourself as an NCIDQ certificate holder on stationery, business cards, and so on, use the following format: "[First name] [Last name], NCIDQ® Certificate No. [######]." For example, "David Ballast, NCIDQ® Certificate No. 9425."

To maintain active status as an NCIDQ certificate holder, you must pay a yearly certificate renewal fee. You will receive your first renewal notice one year after your exam date.

HOW SI UNITS ARE USED IN THIS BOOK

This book uses customary U.S. units (also called English or inch-pound units) as the primary measuring system and includes equivalent measurements in the text and illustrations, using the Système International d'Unités (SI), commonly called the metric system. The use of SI units for construction and publishing in the United States is problematic because the building construction industry (with the exception of federal construction) has generally not adopted metric units. Equivalent measurements of customary U.S. units are usually given as *soft* conversions, whereas customary U.S. measurements are simply converted into SI units using standard conversion factors. Standard conversion results in a number with excessive significant digits. When construction is done using SI units, the building is designed and drawn according to *hard* conversions, where planning dimensions and building products are based on a metric module from the beginning. For example, studs are spaced 400 mm on center to accommodate panel products that are manufactured in standard 1200 mm widths.

As the United States transitions to using SI units, code-writing bodies, federal laws (such as the Americans with Disabilities Act, or ADA), product manufacturers, trade associations, and other construction-related industries typically still use the customary U.S. system and make soft conversions to develop SI equivalents. Some manufacturers produce the same product using both measuring systems. Although there are industry standards for developing SI equivalents, there is no consistent method in use for rounding off conversions. For example, the *International Building Code* (IBC) shows a 152 mm equivalent when a 6 in dimension is required. The *Americans with Disabilities Act and Architectural Barriers Act Accessibility Guidelines* (*ADA/ABA Guidelines*) give a 150 mm equivalent for the same customary U.S. dimension.

To further complicate matters, each publisher may employ a slightly different house style in handling SI equivalents when customary U.S. units are used as the primary measuring system. The confusion is likely to continue until the United States construction industry adopts the SI system completely, precluding the need for dual dimensioning in publishing.

For the purposes of this book, the following conventions have been adopted.

- When dimensions are for informational use, the SI equivalent rounded to the nearest millimeter is used.
- When dimensions relate to planning or design guidelines, the SI equivalent is rounded to the nearest 5 mm for numbers over a few inches and to the nearest 10 mm for numbers over a few feet. When the dimension exceeds several feet, the number is rounded to the nearest 100 mm. For example, if a given activity requires a space about 10 ft wide, the modular, rounded SI equivalent will be given as 3000 mm. More exact conversions are not required.
- When an item is manufactured only to a customary U.S. measurement, the nearest SI equivalent rounded to the nearest millimeter is given, unless the dimension is very small (as for metal gages), in which case a more precise decimal equivalent will be given. Some materials, such as glass, are often manufactured to SI sizes. For example, a nominal $^1/_2$ in thick piece of glass will have an SI equivalent of 13 mm but can be ordered as 12 mm.
- When there is a hard conversion in the industry and an SI equivalent item is manufactured, the hard conversion is given. For example, a 24 in × 24 in ceiling tile would have the hard conversion of 600 mm × 600 mm (instead of 610 mm) because this size is manufactured and available in the United States.
- When an SI conversion is used by a code, such as the IBC, or published in another regulation, such as the *ADA/ABA Guidelines*, the SI equivalents used by the issuing agency are printed in this book. For example, the same 10 ft dimension given previously as 3000 mm for a planning guideline would have a building code SI equivalent of 3048 mm because this is what the IBC requires. The *ADA/ABA Guidelines* generally follow the rounding rule of taking SI dimensions to the nearest 10 mm. For example, a 10 ft requirement for accessibility will be shown as 3050 mm. The code requirements for readers outside the United States may be slightly different.
- Throughout this book, the customary U.S. measurements are given first, and the SI equivalents follow in parentheses. In text, SI units are always given. For example, a dimension might be indicated as 4 ft 8 in (1420 mm). In illustrations, however, standard convention is followed; the SI equivalent is usually without units and is assumed to be in millimeters unless other units are given. The same measurement in an illustration would appear as 4′ 8″ (1420).

CODES, STANDARDS, AND REFERENCES FOR THE EXAM

CODES AND STANDARDS

IDPX covers information related to the following codes and standards.

ADA/ABA Guidelines: *Americans with Disabilities Act and Architectural Barriers Act Accessibility Guidelines*. U.S. Access Board, Washington, DC. www.wbdg.org/ffc/usab/guidelines-standards/ada-aba-guidelines-standards.

BOMA: *BOMA Office Standards Z65.1*, 2017. Building Owners and Managers Association International, Washington, DC.

IBC: *International Building Code*, 2018. International Code Council, Washington, DC.

ICC: *A117.1 Accessible and Usable Buildings and Facilities*, 2009. International Code Council, Washington, DC.

IPC: *International Plumbing Code*, 2018. International Code Council, Washington, DC.

NFPA: *NFPA 101 Life Safety Code*, 2021. National Fire Protection Association, Quincy, MA.

REFERENCES

The following references contain information about the interior design field, including information specific to IDPX content areas. These references may be useful to review as you prepare for the exam.

Allison, Diana. *Estimating and Costing for Interior Designers*. Bloomsbury/Fairchild Books.

Associates III, Kari Foster, Annette Stelmack, and Debbie Hindman. *Sustainable Residential Interiors*. Hoboken, NJ: John Wiley and Sons.

Ballast, David Kent. *Interior Construction and Detailing for Designers and Architects*. PPI.

Ballast, David Kent. *NCIDQ Interior Design Reference Manual*. PPI.

Bonda, Penny, and Katie Sosnowchik. *Sustainable Commercial Interiors*. Hoboken, NJ: John Wiley and Sons.

Botti-Salitsky, Rose Mary. *Programming and Research: Skills and Techniques for Interior Designers, second edition*. Bloomsbury.

Brooker, Graeme, and Weinthal, Lois. *The Handbook of Interior Architecture and Design*. Bloomsbury.

Bush, Pamela McCauley. *Ergonomics: Foundational Principles, Applications, and Technologies*. CRC Press.

Ching, Francis D. K., and Corky Binggeli. *Interior Design Illustrated*. Hoboken, NJ: John Wiley and Sons.

Coleman, Cindy, ed. *Interior Design Handbook of Professional Practice*. New York, NY: McGraw-Hill Professional Publishing.

Crawford, Tad, and Eva Doman Bruck. *Business and Legal Forms for Interior Designers*. New York, NY: Allworth Press.

Downey, Joel, and Patricia K. Gilbert. *Successful Interior Projects Through Effective Contract Documents*. Kingston, MA: R. S. Means Company.

Farren, Carol E. *Planning and Managing Interior Projects*. Kingston, MA: R. S. Means Company.

Godsey, Lisa. *Interior Design Materials and Specifications, third edition*. Bloomsbury/Fairchild Books.

Gordon, Gary. *Interior Lighting for Designers*. Hoboken, NJ: John Wiley and Sons.

Granet, Keith. *The Business of Design: Balancing Creativity and Profitability*. Princeton Architectural Press (Chronical Books).

Harmon, Sharon Koomen, and Katherine E. Kennon. *The Codes Guidebook for Interiors*. Hoboken, NJ: John Wiley and Sons.

Knackstedt, Mary V. *The Interior Design Business Handbook: A Complete Guide to Profitability, fifth edition*. John Wiley & Sons, Inc.

Koe, Frank Theodore. *Fabric for the Designed Interior, second edition*. Bloomsbury/Fairchild Books.

Kopec, Dak. *Environmental Psychology for Design, second edition*. Fairchild Books.

Lechner, Norbert. *Heating, Cooling, Lighting: Sustainable Design Methods for Architects, fourth edition*. John Wiley & Sons, Inc.

McGowan, Maryrose. *Specifying Interiors: A Guide to Construction and FE&E for Commercial Interiors Projects*. Hoboken, NJ: John Wiley and Sons.

McGowan, Maryrose, and Kelsey Kruse, eds. *Interior Graphic Standards*. Hoboken, NJ: John Wiley and Sons.

Mitton, Maureen. *Interior Design Visual Presentation: A Guide to Graphics, Models, and Presentation Techniques, fifth edition*. John Wiley & Sons, Inc.

Mitton, Maureen and Nystuen, Courtney. *Residential Interior Design: A Guide to Planning Spaces, third edition*. John Wiley & Sons, Inc.

Nussbaumer, Linda L. *Human Factors in the Built Environment, second edition*. Bloomsbury/Fairchild Books.

Piotrowski, Christine M. *Professional Practice for Interior Designers*. Hoboken, NJ: John Wiley and Sons.

Reznikoff, S. C. *Interior Graphic and Design Standards*. New York, NY: Whitney Library of Design.

Reznikoff, S. C. *Specifications for Commercial Interiors: Professional Liabilities, Regulations, and Performance Criteria*. New York, NY: Whitney Library of Design.

Rhoads, Marcela Abadi. *Applying the ADA: Designing for the 2010 Americans with Disabilities Act – Standards for Accessible Design in Multiple Building Types*. John Wiley & Sons, Inc.

Slotkis, Susan J. *Foundations of Interior Design, third edition*. Fairchild Books.

Steinfeld, Edward and Maisel, Jordana L. *Universal Design: Creating Inclusive Environments*. John Wiley & Sons, Inc.

Tucker, Lisa M. *Designing Sustainable Residential and Commercial Interiors: Applying Concepts and Practices, first edition*. Fairchild Books.

Tucker, Lisa M. *International Building Codes and Guidelines for Interior Design*. Bloomsbury/Fairchild Books.

Tucker, Lisa M. *Sustainable Building Systems and Construction for Designers, second edition*. Bloomsbury/Fairchild Books.

U.S. Green Building Council. *LEED Reference Guide for Green Interior Design and Construction*. Washington, DC: U.S. Green Building Council.

Wakita, Osamu A. and Linde, Richard M. *The Professional Practice of Architectural Working Drawings, fifth edition*. John Wiley & Sons, Inc.

Wilson, Travis Kelly. *Drafting and Design: Basics for Interior Design*. Fairchild Books.

Winchip, Susan M. *Fundamentals of Lighting, third edition*. Bloomsbury/Fairchild Books.

Winchip, Susan M. *Professional Practice for Interior Designers in the Global Marketplace*. Bloomsbury/Fairchild Books.

Winchip, Susan M. *Sustainable Design for Interior Environments, second edition*. Fairchild Books.

Yates, MaryPaul and Concra, Adrienne. *Textiles for Residential and Commercial Interiors, fifth edition*. Fairchild Books.

PRACTICE QUESTIONS

❶ PROJECT ASSESSMENT AND SUSTAINABILITY

1. The construction fit-out allowance provided to a tenant in an office building would be defined in the

(A) BOMA standards
(B) lease agreement
(C) rental contract
(D) work letter

The answer is D. A work letter is part of the lease agreement between a tenant and the owner of commercial space and defines, among other things, what basic allowances for interior construction the tenant will receive as part of the cost of the lease. These include an allowable number of linear feet of standard partitions, an allowable number of doors and light fixtures, and similar basics required for building out the tenant space. A work letter may also provide for giving the tenant a lump sum amount in lieu of specific quantities of construction elements so the tenant can use the allocated amount, and additional tenant money if required, to build out the tenant's space.

The Building Owners and Managers Association (BOMA) standards give detailed procedures for measuring leased space in commercial buildings. These may be part of the lease agreement, but do not state what allowances are given to a tenant. The lease agreement is the contract between the tenant and the building owner (or manager) and includes rental rates, regulations, other legal details of leasing space, and the work letter. "Rental contract" is not a term normally used in the industry, but such a document would be the same as a lease agreement.

2. An interior designer is calculating the lease area for an office occupying a portion of a high-rise building using the Building Owners and Managers Association (BOMA) standard ANSI/BOMA Z65.1, *Office Buildings: Standard Methods of Measurement.* The exterior wall includes 65% glass area. The designer should calculate the area of the interior space by measuring to the

(A) inside surface of the exterior wall
(B) inside glass surface
(C) outside glass surface
(D) centerline of the exterior wall

The answer is B. According to ANSI/BOMA Z65.1, when the glass is more than 50% of the exterior wall, the measurement is taken to the inside surface of the glass. If the glass is less than 50% of the exterior wall, the measurement should be taken to the inside surface of the exterior wall. The outside glass surface and the centerline of the exterior wall are never used as points of measurement.

3. The use of particleboard as part of a construction detail should be carefully evaluated in terms of the particleboard's

(A) availability
(B) permeability
(C) strength
(D) VOC content

The answer is D. Most particleboard contains formaldehyde and may contain other volatile organic compounds (VOCs). The other options are incorrect because particleboard is readily available and strong enough for architectural woodwork details and other uses. Permeability, which is the ability to transfer moisture, is not a factor because the particleboard would only be used as a substrate for other finishes.

4. To determine the necessary rentable floor area for a project, the interior designer must know which of the following? (Choose the three that apply.)

(A) net assignable area
(B) number of employees
(C) efficiency factor
(D) expected growth rate
(E) corridor size and configuration
(F) rentable-usable ratio

The answer is A, C, and F. Once the interior designer has determined the net assignable area, this value is divided by the efficiency factor, which takes into account nonusable spaces (such as corridors) to calculate the usable area. The rentable-usable ratio (or load factor) is multiplied by the usable area to account for common building spaces as established by the building owner. Although the number of employees does affect the required net assignable area, it does not determine the required floor area. If the expected growth rate needs to be accounted for, it should already be factored into the net assignable area. The corridor size is reflected in the efficiency factor, and corridor configuration would be determined at a later date.

5. The initial determination of the area required for a client's program gives the

(A) gross area
(B) net area
(C) rentable area
(D) usable area

The answer is B. Regardless of how the information is collected, area is based on the actual space that a client needs to perform a function. This is the net area or the net assignable area. For example, a client would know that a 150 ft^2 (14 m^2) office is required but would not give consideration to the corridor required to get to the space or the wall thickness needed to create the office. Based on the net area and knowledge of the project type, the programmer can estimate how much additional space is required for secondary circulation. The usable space can then be used as a basis for calculating the rentable and, if necessary, the gross area.

6. An interior designer has been hired to redesign a large Victorian house into a bed and breakfast inn. During an initial interview, the client says that he wants to restore the interior to an authentic Victorian look, as well as enlarge some of the bedrooms into suites so that the inn will be the best in town. After hearing these comments, what should be the designer's FIRST course of action?

(A) Suggest that the client also retain an architect to determine the feasibility of enlarging rooms and removing walls.
(B) Ask the client if he has a budget, and suggest conducting a preliminary cost estimate to see if he can afford what he wants.
(C) Tell the client that he needs to define what he means by "the best in town," to give the designer a more definite idea of how to proceed.
(D) Recommend to the client that field measurements of the house be conducted, and begin research on authentic Victorian furnishings and finishes.

The answer is B. Feasibility should be determined before any other work is done. In a case like this, the client is not likely to have a good grasp of the costs required for remodeling and may not have enough of a budget to do the job as he wants to. Option A is the second-best choice, but would require additional costs to the client even before the feasibility was determined. Options C and D are important issues but should be considered only after the overall feasibility of the project has been determined.

7. Who is responsible for determining the condition of an existing floor as part of a due diligence investigation?

(A) flooring subcontractor
(B) general contractor
(C) interior designer
(D) structural engineer

The answer is C. The due diligence investigation of an existing floor involves verifying the general floor condition and generally requires no special expertise. Therefore, the interior designer is responsible for it, and should be the one to call in other experts if necessary. However, such aspects as the structural capacity of the floor must be determined by a structural engineer.

8. Due diligence requires the investigation and documentation of which of the following? (Choose the four that apply.)

(A) client's requirements for specialized spaces
(B) adequacy of the existing mechanical and electrical systems
(C) accuracy of any existing drawings of the space
(D) structural capacity and existing means of egress in the building
(E) elements related to the condition of the existing space
(F) types of adjacent tenants and the spaces they occupy

The answer is B, C, D, and E. Due diligence requires the investigation, understanding, and documentation of a space and its surrounding context. However, the client's programming requirements and determining the specific types of adjacent tenants are not part of the process.

9. A client has requested a new entry to her consulting business, which is located in an old, nonsprinklered building. The client's space must conform to current *International Building Code* (IBC) requirements. The current entrance consists of a pair of all-glass doors mounted on floor closers. The entrance opens onto a 1-hour-rated building corridor. What should the designer tell the client to expect regarding the new entrance?

(A) Smoke seals will have to be located around the edges of the glass doors.
(B) The glass doors will have to be replaced.
(C) One of the doors will have to be removed.
(D) The floor closers will have to be changed to hinges.

The answer is B. Smoke seals will be required, but the most important thing is that the glass doors will have to be replaced, either with solid, 20-minute-rated doors or with 20-minute-rated doors with glass that is also 20-minute rated. Because this will significantly change the appearance of the existing entry, it is the first thing the client should be told to expect. Either pivoted floor closers or hinges may be used as long as they are also fire-rated and the door is side-swinging.

10. A project manager is completing the due diligence site analysis for a building to be used by a law firm. Before beginning space planning, the project manager is MOST likely to need expert consulting assistance to determine the

(A) type of in-floor electrical and communications outlets
(B) number and locations of sprinkler heads
(C) reinforcement needed to accommodate a new law library
(D) amount of additional natural lighting needed for daylighting

The answer is C. The additional dead load of a law library could be significant enough to require extra structural reinforcement. The project manager would require the assistance of a structural engineer to determine this.

Without the assistance of a consultant, the project manager could determine the number and locations of sprinkler heads, whether the natural lighting is sufficient for daylighting, and the type of in-floor electrical and communications outlets that are needed. (However, the capacity of the electrical system for additional or unusual loading would have to be determined by an electrical engineer.)

11. The building commissioning of an interior design project seeking LEED credit is the responsibility of the

(A) building architect
(B) independent commissioning team
(C) project's interior designer
(D) project's mechanical engineer

The answer is B. In order to get LEED credit, an independent commissioning team that does not include anyone responsible for the project must be used. Even without seeking LEED credit, commissioning requires a joint effort of the mechanical engineer, the contractor, the electrical engineer, the building owner, and others.

12. An interior designer is beginning work on an extensive residential remodeling project of a house that was built in 1970. Which of the following materials is of most concern to investigate the presence of? (Choose the four that apply.)

(A) asbestos
(B) lead
(C) PCBs
(D) radon
(E) vermiculite
(F) vinyl trim

The answer is A, B, D, and E. Of the four choices, lead would probably be the most problematic. Lead is commonly found in paint in homes built before 1978. As lead can cause serious health problems, especially to children, the designer should tell the owner that the existing paint should be tested. If lead is discovered, it must be removed by certified removal specialists. Asbestos could also be present in asphalt flooring, insulation, and textured paints. Radon in the soil and vermiculite in insulation could also be present, and these substances should be investigated.

Polychlorinated biphenyls, or PCBs, are more likely to be found in commercial and industrial applications. Vinyl in trim and other building projects, while not the best choice for sustainability, is not dangerous in itself as a finished product.

13. When the interior designer helps the client establish goals and objectives, these normally lead to which of the following?

(A) a determination of the areas required for spaces in the project
(B) a succinct statement of the problem
(C) assistance with separating "wants" from "needs"
(D) the establishment of programmatic concepts

The answer is D. Goals and objectives are simply what the client wants to achieve in the project. They may be very general or vague or may be very specific. The designer must then develop these goals and objectives into programmatic concepts that can lead to design concepts that suggest the physical means of achieving those goals. The determination of spatial needs, a final set of problem statements, and wants versus needs, come later in the programming process.

14. Locating and designing a stairway in a building to encourage everyday use is an example of

(A) active design
(B) design for wellness
(C) resilient design
(D) sustainable design

The answer is A. Active design is a set of design principles that promote physical activity. Designing a stairway that is conveniently located, near building entrances or near elevators, and comfortable to use encourages people to get more physical activity, thus promoting health.

Wellness design is the planning and design of environments with socially conscious systems and materials to promote the well-being of people, including physical, emotional, and cognitive states. Resilient design is the design of buildings, communities, and regions to help adapt and respond to natural and manmade disturbances. Sustainable design is meeting the needs of the present without compromising the ability of future generations to meet their own needs, including wise use of materials, energy conservation, alternate energy sources, indoor air quality, and recycling.

15. In the LEED v4 for Interior Design and Construction rating system, the minimum credit of two points for indoor water use reduction for commercial interiors can be achieved by which of the following goals regarding bath, lavatory, and kitchen fixtures?

(A) specify fixtures that meet the baseline water consumption values
(B) reduce aggregate water consumption by 20% from the baseline values
(C) reduce water use from the baseline by 25%
(D) reduce water use from the baseline by 35%

The answer is C. To achieve the minimum credit of two points for commercial interiors the reduction above the prerequisite values must be at least 25%. A 35% reduction is worth six points. The baseline water consumption values give the starting point for calculating reductions. A value of 20% from baseline values is just the prerequisite and is not worth any points.

❷ PROJECT PROCESS, ROLES, AND COORDINATION

16. Which item in the following excerpt from a specification is a performance specification?

Part 2—Products

2.01 Metal Support Material

General: To the extent not otherwise indicated, comply with ASTM C754 for metal system supporting gypsum wallboard.

Ceiling suspension main runners: $1^1/_2$ in steel channels, cold rolled.

Hanger wire: ASTM A641, soft, Class 1 galvanized, prestretched; sized in accordance with ASTM C754.

Hanger anchorage devices: Size for 3 times calculated loads, except size direct-pull concrete inserts for 5 times calculated loads.

Studs: ASTM C645; 25-gage, $2^1/_2$ in deep, except as otherwise indicated.

Runners: Match studs; type recommended by stud manufacturer for vertical abutment of drywall work at other work.

(A) ceiling suspension main runners
(B) hanger wire
(C) hanger anchorage devices
(D) runners

The answer is C. The hanger anchorage device is a performance specification. The ceiling suspension main runners are a descriptive specification. The hangar wire is a reference standard specification. The runners are also a reference standard specification because the wording refers to the reference standards of the studs.

17. The "latest possible starting time" is found in which of the following schedules or charts?

(A) CPM chart
(B) full-wall schedule
(C) Gantt chart
(D) workflow schedule

The answer is A. A critical path method (CPM) chart graphically depicts all the tasks required to complete a project, the sequence in which they must occur, and their duration. As the name implies, the chart determines which tasks and durations are the most critical to be completed on schedule if the entire project is to be completed on time. Other tasks may start and finish a little earlier or later without affecting the final completion date. Thus, a task could have a "latest possible starting time" attached to it.

A full-wall schedule is a manual way of scheduling using 3 in × 5 in cards on a wall; each card lists needed tasks and includes a timeline across the top and names of people completing the tasks along the left side. A Gantt chart is another name for a bar chart; it uses horizontal bars to indicate the duration of tasks coordinated, and includes a timeline across the top and the individual tasks listed along the left side. A workflow schedule is not a type of chart used in the interior design profession.

18. Which of the following provides the MOST accurate basis for estimating project costs?

(A) square footage
(B) functional unit
(C) parameter
(D) quantity takeoff

The answer is D. A quantity takeoff is the most detailed method of developing a budget and, therefore, it is the most accurate. To use this method, count the actual quantities of materials and furnishings and multiply these quantities by firm, quoted costs.

19. The MOST important source of information for monitoring project budgeting during design and development of construction documents is the

(A) aged accounts receivable report
(B) fee projection chart
(C) project progress report
(D) personnel timesheet

The answer is D. Each individual staff member's timesheet provides the basis for developing the project progress report (or a simple manually produced graph) that is used for monitoring a project. The aged accounts receivable report shows the status of all invoices for all projects and whether they have been paid or not. It has nothing to do with monitoring an individual project for comparison with the budgeted time and fees. A fee projection chart is necessary for comparing actual time and fees expended, but by itself is not a basic source of information. A project progress report does compare actual time and fees with budgeted time and fees for a project, but it is not the *source* of the information.

20. When an interior designer is coordinating with an acoustical consultant, the designer's drawings should show

(A) copies of the consultant's sound control diagrams
(B) a schedule of STC ratings for all partitions
(C) the consultant's specifications for equipment
(D) construction details of critical acoustical elements

The answer is D. Acoustical consultants rarely produce separate drawings like those of an electrical or a mechanical engineer. However, they do suggest details and diagrams for design and construction, as well as specifications for materials that are critical for sound and noise control. These details and diagrams are redrawn (not copied) by the interior designer and placed on the designer's drawings (or, if they are CAD details, they are reproduced directly). Schedules of sound transmission class (STC) ratings are not used. Any specifications the consultant provides are placed in the project manual and not on the drawings.

21. A CPM chart

(A) takes less time to develop than a Gantt chart
(B) is simpler than a PERT chart
(C) offers a simplified view of the project schedule
(D) shows the shortest possible project schedule

The answer is D. A critical path method (CPM) chart shows the path through the network of activities that will result in the shortest possible project schedule. It gives the earliest and latest possible start and finish dates for the various activities so that the project stays on schedule. A CPM chart presents a complex network of interconnected nodes and lines of activities and durations, making it almost as complex as a programming evaluation and review technique (PERT) chart. A CPM chart takes more time to develop than a Gantt or bar chart.

22. An interior designer is teaming with an architect and working under the architect's contract with the owner to complete a new hotel project. The owner for the project will be using a construction manager as adviser (CMa) as part of the project team using AIA Document A232, *General Conditions of the Contract for Construction, Construction Manager as Adviser Edition*. What assistance can the interior designer expect the construction manager to provide to the interior designer?

(A) recommendations for selection of contractors and furniture dealers
(B) advice on constructability, cost estimating, and project scheduling
(C) assistance with fast-track design procedures and construction methods
(D) participation in construction administration and project closeout

The answer is B. The interior designer can expect the construction manager to provide early advice on constructability, cost, and scheduling. A CMa works with the owner and architect in selecting contractors, but does not recommend furniture dealers. While a CMa does assist with fast-track construction, there is no mention of using this delivery method in the question. The CMa does participate in construction administration and project closeout, but in this scenario the architect would be involved with these activities under the A232 document, with the interior designer as a secondary participant.

23. Before preparing construction documents, the interior designer should verify whether any particular types of lighting fixtures are required by coordinating with the

(A) architect
(B) building owner
(C) contractor
(D) electrical engineer

The answer is B. The building owner may specify the types of light fixtures to be used in the building. If so, this specification should be stated in the contract documents. Therefore, the interior designer should check with the building owner before preparing the documents.

24. Which chart is MOST appropriate for a short, simple project?

(A) CPM chart
(B) Gantt chart
(C) line chart
(D) PERT chart

The answer is B. A Gantt chart (also called a bar chart) is a simple representation of a schedule, and it works well for projects of short duration that do not require complex coordination. Each bar represents an activity and is plotted along a horizontal axis representing time. The duration of each activity and how activities overlap or depend on one another can be shown easily.

A critical path method (CPM) chart is a network diagram that is used to show all the tasks that need to be completed, when the tasks need to start, and when the tasks will finish. From the chart, a critical path can be established that represents the shortest possible schedule, as well as the dates on which critical tasks must start, and if the schedule will be met. A programming evaluation and review technique (PERT) chart is similar to a CPM chart, but uses different networking methods. Both CPM and PERT charts are overly complex for simple schedules. A line chart is not a type of scheduling chart; it is a listing of the products that a manufacturer's representative carries.

25. To obtain a building permit after the contract documents are completed, the interior designer should

(A) give the documents to the client, who will transmit them to the authority having jurisdiction (AHJ)
(B) forward the required number of drawing sets to the contractor, who will submit them to the AHJ
(C) submit the documents directly to the AHJ and provide a copy to the contractor
(D) provide the documents when requested by the AHJ after submitting a permit application and fee

The answer is B. The contractor is typically responsible for submitting the construction documents, along with an application and fee, to the building department or AHJ.

26. What type of specification does the following excerpt represent?

PART 2 PRODUCTS

2.01 Materials

A. Paints shall be as manufactured by ICI Dulux Paints. Colors, types, and sheens shall be as listed on the finish legend on the drawings. Formulation of primer and finish coats shall be selected by the painting contractor based on the type of substrate being covered and the specified sheen.

B. Paints shall comply with applicable federal, state, and local laws enacted to insure compliance with Federal Clean Air Standards and shall conform to the restrictions of the local air pollution control authority.

(A) descriptive
(B) or equal
(C) performance
(D) proprietary

The answer is D. This specification names a specific manufacturer with reference to specific colors and types on the finish legend. This is a proprietary specification.

27. In preparing a budget, which of the following elements would an interior designer MOST likely ask the client to include?

(A) legal fees and artwork consulting
(B) taxes on furniture and computer system installation
(C) contingencies and an interior designer's fee
(D) furnishings and furniture delivery

The answer is A. Legal fees and specialty consulting (like artwork advice) are often separated from the construction and furnishing budget that the interior designer prepares. The items in the other three options can be estimated by the interior designer (unlike legal fees) and are often placed in the designer's budget work.

28. What type of specification does the following excerpt indicate?

C. Sheet vinyl flooring:

1. Sheet vinyl flooring shall conform to ASTM F 1303, Type I [Type II], Grade 1 [Grade 2] [Grade 3], backing Class A [B] [C]. [Provide manufacturer's standard welding rods for heat welding of joints.]

Select from below. If a seamless installation is desired, select the color of heat welding rod you want from the manufacturer's standard.

2. Color shall be as shown on the finish schedule. [_____________.] [Color of heat welding rods shall be as shown on the finish schedule] [A specific manufacturer's color is for reference purposes. Other products from approved manufacturers listed above will be considered in accordance with the "Substitution" provisions in Section 01300 and in the Instructions to Bidders.]

D. Rubber tile flooring

Rubber flooring is good for slip resistance and it reduces leg fatigue. However, it is not resistant to oil and grease and some types of staining.

(A) descriptive specification
(B) guide specification
(C) master specification
(D) reference standard specification

The answer is C. The presence of selections from which to choose and notes to the specifier (shown in bold) indicate that this is a master specification. A guide specification would not be as complete and would not show notes to the specifier, just the overall organization, paragraph selections, and blanks to be filled in by the specifier. Descriptive and reference standard specifications are specific types that contain final specification language with no choices or guide notes.

29. Requirements for vertical louver blinds would be found in which part of the CSI MasterFormat outline?

(A) Division 07
(B) Division 09
(C) Division 10
(D) Division 12

The answer is D. Vertical louver blinds are specified in Division 12, "Furnishings," under section 12 20 00, "Window Treatments." Division 07 is "Thermal and Moisture Protection," Division 09 is "Finishes," and Division 10 is "Specialties."

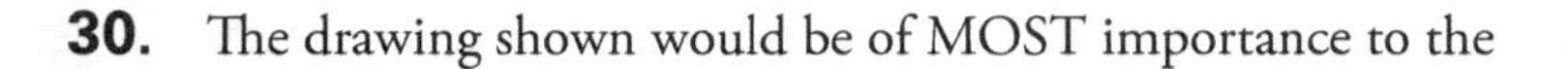

30. The drawing shown would be of MOST importance to the

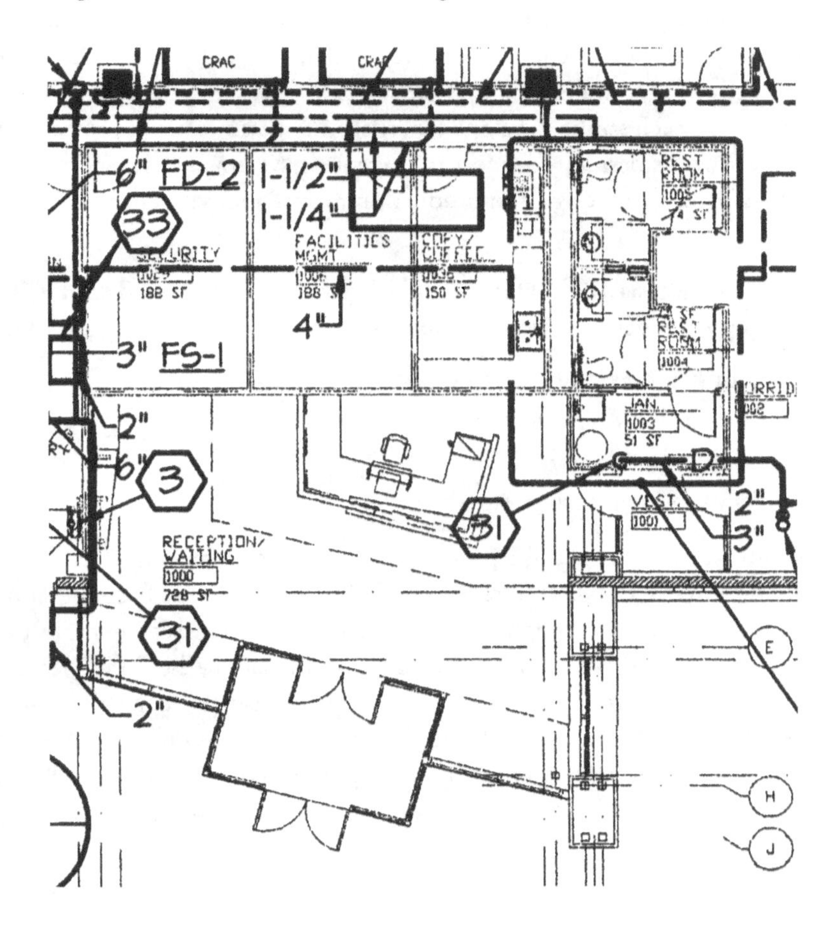

(A) general contractor
(B) plumbing subcontractor
(C) mechanical subcontractor
(D) fire protection subcontractor

The answer is B. This partial drawing is a plumbing plan showing the size and routing of piping for the restrooms and other plumbing fixtures. This is what the plumbing subcontractor uses to install plumbing.

❸ PROFESSIONAL BUSINESS PRACTICES

31. A medium-size design firm specializes in three different project types, and uses a project management system where project managers have daily contact with design and production teams, as well as with clients. The firm would BEST be set up using a

(A) departmental organization
(B) project manager organization
(C) strong principal organization
(D) studio organization

The answer is D. The firm would be best set up using a studio organization. A studio organization assigns smaller groups of employees (called studios) to different project types, such as commercial, residential, or retail projects. Each studio is then responsible for completing all project phases, from initial programming to closeout. In this way, each studio develops expertise in a specific type of project. A studio organization works best with project management systems where project managers have daily contact with design and production teams, and with clients.

32. A client requires a 100,000 ft^2 (9290 m^2) project be completed as quickly as possible. Which project delivery method should the interior designer suggest to the client?

(A) construction manager method
(B) design-bid-build method
(C) design-build method
(D) multiple contracts method

The answer is A. For large projects that can absorb the additional overhead, a construction manager (also abbreviated "CM") can advise on cost and constructability questions during the design process and, afterward, can set the project up on a fast-track schedule. Although using the design-bid-build method often results in the lowest project cost through competitive bidding, it is slower than the construction manager method. The design-build method is intended to offer the client a fixed price and single-source responsibility, not the quickest construction completion time. The multiple contracts method also usually increases construction time.

33. Which of the following has the MOST influence on the scheduling of a design project?

(A) time required for regulatory review
(B) number of design office staff
(C) size and complexity of the project
(D) design methodology of the project team

The answer is C. In most cases, the size and complexity of a project have the most influence on the scheduling of a design project. The number of design office staff has less influence on the scheduling, and even with a large staff, there is a point beyond which more people will not move the project along more quickly. The time required for regulatory review usually does not vary, and the rest of the schedule is set to work around it. The method of designing does not have as much influence on scheduling as do the size and complexity of the project.

34. Which of the following are typically required of an interior design firm that resells furniture and bills clients?

(A) local business license and corporate identification number
(B) corporate identification number and resale license
(C) local business license and sales tax license
(D) corporate identification number and sales tax license

The answer is C. Interior design firms can usually conduct business (and bill clients) with only a local business license. However, if a firm plans to resell furniture, it will also need a sales tax license (also called a resale license or transaction privilege tax license). State agencies issue corporate identification numbers to corporations only, not to design firms.

35. After construction has begun, an owner has a friend add a faux finish to two rooms. Which course of action must be followed if the general contractor is working under AIA Document A201, *General Conditions of the Contract for Construction*?

(A) The owner must tell the general contractor that additional work will be done under a separate contract. The friend must then coordinate with the general contractor as necessary.

(B) The owner must cancel the initial contract, renegotiate with the general contractor, and write a new contract to include the friend and new work.

(C) The owner and friend must negotiate and write an agreement separate from that of the general contractor. The agreement must require that the work will not interfere with the general contractor's primary work.

(D) The owner must wait until the general contractor has completed the primary work before having the faux finish contractor start work.

The answer is A. Under AIA Document A201, *General Conditions of the Contract for Construction*, the owner may contract work separately from the contract with the general contractor. However, the owner and additional contracted workers must coordinate and cooperate with the general contractor.

36. A standard American Institute of Architects (AIA) contract between an interior designer and a client includes which of the following? (Choose the four that apply.)

(A) furniture vendor responsibilities

(B) project cost estimating

(C) agreement termination methods

(D) method of producing the drawings

(E) client responsibilities

(F) method of dispute resolution

The answer is B, C, E, and F. The interior designer's compensation, procedures for termination of the contract or agreement, the responsibilities of the client, and acceptable methods of dispute resolution are critical parts of agreements and contracts, and are included in the standard AIA contracts between interior designers and clients. Furniture vendor responsibilities are not part of standard contracts. The standard AIA contracts define the instruments of service and their use, but do not specify how they are to be produced.

37. During the preparation of a construction contract, the client asks the interior designer to supply the correct amounts of coverage to be included in the *Supplementary General Conditions of the Contract.* The designer should

(A) check with an insurance agent before supplying the amounts

(B) mention that insurance amounts are not necessary because they are covered in the *General Conditions of the Contract*

(C) set up a meeting with the client and an insurance agent to go over insurance amounts together

(D) remind the client that it is the client's responsibility to obtain insurance information

The answer is D. AIA Document B152, *Standard Form of Agreement Between Owner and Architect for Interior Design and Furniture, Furnishings, and Equipment (FF&E) Design Services,* clearly states that it is the client's responsibility to provide all legal, accounting, and insurance information for the project. The interior designer is not responsible for obtaining or providing these types of information.

38. An interior designer has contracted with the owner of a large commercial building to do tenant finish work for new building occupants. The work will consist of space planning for tenants that lease building space, and construction drawings that the building's contractor will use to build out each space. What is the BEST way for the interior designer to charge the building owner?

(A) area fee

(B) fixed fee

(C) hourly fee

(D) cost plus fee

The answer is A. An area fee is a professional fee based on multiplying the square footage of a project by some fixed rate. Tenant finish work is often priced on a square-foot basis because the interior designer knows how much work will be required for each tenant, and because most tenants, building owners, and leasing agents negotiate such work on a square-foot basis. However, as with any fixed fee, the scope of services must be clear so that the interior designer does not do more work than the fee allows.

39. What type of insurance protects the designer in case a mistake on a drawing results in bodily injury or property damage?

(A) general liability

(B) personal injury

(C) professional liability

(D) workers' compensation

The answer is C. Professional liability insurance (also called errors and omissions insurance or malpractice insurance) protects a designer who makes a professional error, such as a mistake on a drawing or a specification, that results in bodily injury or property damage.

40. An interior designer wants to reduce fees to be competitive in a local market. The designer should first consider reducing

(A) indirect labor costs
(B) capital expenses
(C) profit
(D) salary rates

The answer is A. Indirect labor accounts for the largest percentage of overhead costs in a design firm; therefore, it should be reduced first. Option B is incorrect because capital expenses typically do not account for a large percentage of overhead costs and they can be deferred to a later time if necessary. Although interior designers often reduce their percentage of profit, profit should remain the same in order to keep the business viable. Therefore, option C is incorrect. Option D is incorrect because salaries are not considered overhead. In addition, salaries are set at the time of employee hiring, and thus, cutting them generally is not an option (or at least is not a professional one) for most professional businesses.

❹ CODE REQUIREMENTS, LAWS, STANDARDS, AND REGULATIONS

41. A designer is planning a library in which tall bookshelves will be used. If the project is located in a city that has adopted the *International Building Code* (IBC), where would the designer look to find requirements on the minimum allowable space between the top of the shelving and the sprinkler heads in the ceiling?

(A) IBC
(B) *International Mechanical Code*
(C) NFPA 13
(D) IPC

The answer is C. The IBC refers to NFPA 13 in detailing the requirements of sprinkler system design and installation. The other model codes refer to NFPA 13 as well.

42. The following abbreviated table includes requirements for occupancy loads. A restaurant on the ground floor contains 3500 ft^2 (325 m^2) of dining area, a 1000 ft^2 (93 m^2) kitchen, and a 1200 ft^2 (111 m^2) bar. What is the total occupant load?

use	occupant load factor (ft^2/occupant)
assembly areas, concentrated use (without fixed seats) auditoriums, dance floors, lodge rooms	7
assembly areas, less-concentrated use conference rooms, dining rooms, drinking establishments, exhibit rooms, lounges, stages	15
hotels and apartments	200
kitchens, commercial	200
offices	100
stores, ground floor	30

(A) 202 occupants

(B) 318 occupants

(C) 380 occupants

(D) 410 occupants

The answer is B. From the table, assembly areas including restaurants and bars have an occupant load of 15 ft^2 (1.39 m^2) per occupant. Commercial kitchens have an occupant load of 200 ft^2 (18.6 m^2) per occupant. Therefore,

$$\frac{3500\ \text{ft}^2}{15\ \frac{\text{ft}^2}{\text{occupant}}} = 233\ \text{occupants}$$

$$\frac{1000\ \text{ft}^2}{200\ \frac{\text{ft}^2}{\text{occupant}}} = 5\ \text{occupants}$$

$$\frac{1200\ \text{ft}^2}{15\ \frac{\text{ft}^2}{\text{occupant}}} = 80\ \text{occupants}$$

$$\text{total} = 233\ \text{occupants} + 5\ \text{occupants} + 80\ \text{occupants} = 318\ \text{occupants}$$

43. Which of the following are the MOST effective strategies for maximizing indoor air quality? (Choose the four that apply.)

(A) conducting a post-occupancy evaluation
(B) developing a maintenance manual
(C) maximizing recirculation of filtered indoor air
(D) planning separate rooms for large copiers
(E) specifying materials with low VOCs
(F) prohibiting smoking on all floors of the building

The answer is B, D, E, and F. Developing a maintenance manual, using separate rooms for large copiers, specifying materials with low volatile organic compounds (VOCs), and prohibiting smoking on all floors are the best ways to establish and maintain good indoor air quality.

Although post-occupancy evaluation is a good tool for verifying that earlier decisions regarding indoor air quality are being maintained, it does little for actually maximizing indoor air quality. Instead of maximizing recirculation of indoor air, outdoor air should be brought into the building.

44. An interior designer has been retained to redesign a call center in a large warehouse. The call center will have a high-density employee occupancy. As part of due diligence, what must the interior designer do to determine whether the existing sanitary fixtures will be sufficient?

(A) Count the number of existing sanitary fixtures and compare it to the requirements of the AHJ.
(B) Ask the client how many fixtures were previously used and document this number in the due diligence report.
(C) Find the required number of fixtures in the *International Plumbing Code* and use this full number in the new design.
(D) Ask the local building official for a recommendation.

The answer is A. The *International Plumbing Code*, as amended by the local authority having jurisdiction (AHJ), will give the minimum required number of sanitary fixtures (e.g., toilets, urinals, lavatories, and drinking fountains) based on the occupancy type and occupant load (i.e., the number of people to use the fixtures). The interior designer can count the number of existing fixtures, make the comparison, and tell the client what the planning and cost implications are. If questions remain or if it is still unclear how many fixtures are required, the interior designer can then ask the local building official, though doing so is not always necessary nor part of due diligence. Also, the building official would only know the total number of fixtures required, not the existing number of fixtures. It would still be necessary for the designer to count the existing fixtures to determine whether the number sufficiently meets code and AHJ requirements. Including the full number of code-required fixtures in the design is unnecessary and costlier than just including the difference between what is needed and what already exists. The client is not qualified to determine the minimum number required by codes or regulations.

45. A manufacturer claims that its product meets an ASTM standard, but refers to the standard only by number and title. The interior designer can BEST review the standard by

(A) calling the trade association representing the type of product in question
(B) getting the standard from the manufacturer
(C) ordering the standard directly from ASTM or from a third-party seller
(D) searching a university or research library

The answer is C. ASTM International sells the standards it develops, and they are available both in print and online for a fee. Third-party sellers such as Techstreet, IHS, and NSSN also have the standards for sale. Manufacturers may have the standards related to their products, and trade associations may have those related to the kinds of products they represent, but in most cases they will not offer them to architects and designers. Some libraries have collections of major standards from ASTM and other trade organizations, but these collections usually are not current and do not include most of the specialty standards of the interior design and architectural professions.

46. The interior designer can find the requirements for the maximum flow rates for plumbing fixtures for a project by consulting which of the following?

(A) authority having jurisdiction (AHJ)
(B) *International Green Construction Code*
(C) *International Plumbing Code*
(D) WaterSense Program of the Environmental Protection Agency (EPA)

The answer is A. To determine the maximum flow rates for a specific project, the interior designer must obtain this information from the local AHJ, as many states and local jurisdictions have established requirements that differ from those in other codes and standards, including the *International Plumbing Code.*

The *International Green Construction Code* provides both mandatory and voluntary standards for water use, but applies only if it has been adopted by the local jurisdiction. The *International Plumbing Code* gives maximum flow rates for various types of plumbing fixtures, but can be superseded by the local AHJ. The WaterSense program is a voluntary program that certifies products that use 20% less water than the federal minimum established by the U.S. Energy Policy Act of 1992 without sacrificing performance.

47. An interior designer's client has purchased a small, low-hazard, sprinklered, warehouse in an industrial section of a city and is considering turning it into a small concert venue with a restaurant and bar. What aspects of the change in use should the interior designer first be most concerned with?

(A) if the building code will require an upgraded fire protection system
(B) the smaller amount of square footage of the warehouse that will be allowed for the new use
(C) whether the local zoning ordinance allows the change in use
(D) if the concert venue and restaurant/bar can be in the same building

The answer is B. For the same construction type and sprinklering provisions, the *International Building Code* (IBC) restricts assembly occupancies to a smaller area than a low-hazard warehouse. The designer would have to verify that this was acceptable to the client or propose ways to increase the building type rating, separate the building into separate fire areas, or use some other method for satisfying the program.

The fire protection system may have to be modified, but this could be verified at a later time; it wouldn't be the first issue to deal with. A concert venue and restaurant would be higher zoning classifications than a warehouse, so it is likely that these changes in the use would be allowed. The opposite would not be true—trying to turn a restaurant and concert hall into a warehouse. Two uses in the same building would be considered a mixed occupancy and would be allowed as long as building code requirements for separation with fire-rated construction were created.

48. One of the best ways for an interior designer to implement water conservation is to

(A) plan residential uses for only showers instead of bathtubs and showers
(B) investigate the possible use of graywater for some applications
(C) reduce the number of plumbing fixtures in a building
(D) specify fixtures that use less water than LEED or IPC maximums

The answer is D. The simplest way to reduce water use is to use less of it. While Leadership in Energy and Environmental Design (LEED) requirements and those in the *International Plumbing Code* (IPC) give maximum flow rates for various types of fixtures, it is possible to find fixtures that use less, for example, waterless urinals.

For all residential occupancies, either showers or bathtubs are required; so specifying only a shower is possible, but the decision should be made with the client as the client may prefer to have a bathtub. There are very few jurisdictions that allow the use of graywater, and it must be plumbed separately from potable water; so even if it was possible, it would be an expensive option. Reducing the number of fixtures is not possible as the minimum number is set by the IPC and the *International Building Code* (IBC).

49. An interior designer is working on a retrofit project for a building that was designed using compact fluorescent lamps and incandescent downlights. What action could the designer take to improve the energy conservation of the buildings?

(A) replace all the lights with LEDs
(B) change the incandescent downlights to compact fluorescents
(C) reduce the number of lamps in each luminaire where possible
(D) replace the downlights with compact fluorescent troffers

The answer is A. The best and most flexible way to reduce electrical use is to use more efficient LEDs (light emitting diodes) in all the fixtures. Changing incandescents to fluorescents would help but not as much as LEDs. Reducing the number of lamps would severely limit the amount of light unless they were already over-lamped to begin with. Replacing downlights with troffers would probably negate the focused direction of the light as was originally intended.

50. Good indoor air quality in an office space is important for which of the following reasons? (Choose the four that apply.)

(A) reduces absenteeism
(B) improves a person's sense of well-being
(C) prevents mold
(D) maintains health
(E) improves productivity
(F) controls humidity

The answer is A, B, D, and E. Good indoor air quality in an office space can reduce absenteeism, improve a sense of well-being, maintain health, and improve productivity, as well as improving creativity and motivation.

Mold is caused by microscopic organisms that require moisture, a nutrient, and a proper temperature range in which to grow. These can be present even in a space with good indoor air quality. Good indoor air quality by itself does not minimize humidity; a proper HVAC system is required to do that.

51. When evaluating building materials for their sustainability potential, which of the following should be considered? (Choose the four that apply.)

(A) potential for reuse and recycling
(B) fire resistance
(C) durability of materials
(D) low toxicity
(E) material availability
(F) embodied energy

The answer is A, C, D and F. The potential for reuse and recycling, durability, low toxicity, and embodied energy are all aspects of material sustainability potential that should be reviewed when selecting materials.

Fire resistance is not a concern with sustainability. Any material is either fire resistance or not whether it has a sustainability potential. Material availability does not necessarily imply sustainability. While materials sourced close to the construction site may qualify for LEED credit it does not necessarily mean it is truly sustainable.

52. Which of the following materials has the greatest potential for reuse in interior construction?

(A) aluminum
(B) plastics
(C) steel
(D) wood

The answer is C. Both steel and aluminum can be melted and reformed into other construction products. Steel is best with a potential recycled content of 30% or more. Aluminum may have a recycled content of 20% or more.

Some types of plastics salvaged from a building can be ground and used as material for new products, but for interior construction this is difficult. Recycling wood can also be problematic. Untreated wood can be turned into particleboard or other engineered wood products. However, treated or painted wood generally cannot be recycled into building products.

53. Zoning maps indicate which of the following?

(A) front yards
(B) off-street parking
(C) allowable uses
(D) building setbacks

The answer is C. A zoning map indicates the borders of zoning districts, typically with an alpha-numeric numbering system. The numbers relate to the zoning code itself, which gives the allowable uses for each zoning district. The requirements for front yards, off-street parking, and building setbacks are not given on the zoning map itself but in the zoning code.

54. An interior designer is working on a remodeling project for a family with three small children. The house, originally built in 1971, has four bedrooms and a large family room. What potential contaminant should the designer advise the client to test for?

(A) formaldehyde
(B) lead
(C) PCB
(D) VOCs

The answer is B. The most critical hazardous material that should be tested for is lead. Lead is a highly toxic metal that was once used in a variety of consumer and industrial products. Exposure to lead can cause serious health problems, especially to children. Most exposure to lead comes from paint in homes built before 1978.

Formaldehyde is inherent in many building products, so it really cannot be removed without removing the product it is a part of. It is best mitigated by the interior designer by specifying materials that do not use it or that include it in acceptable amounts. Polychlorinated biphenyl, or PCBs, are used mainly in commercial and industrial applications. As with formaldehyde, volatile organic compounds (VOCs) are within a product, so the interior designer should specify materials that have no VOCs or those that meet acceptable limits.

55. Which of the following can the interior designer implement as part of the designer's work product for energy conservation in a new building? (Choose the four that apply.)

(A) specify building commissioning
(B) design a displacement ventilation system
(C) specify automatic occupancy lighting controls
(D) design and finish occupied spaces near windows with daylight-responsive controls
(E) use LEDs throughout the space
(F) require a variable refrigerant flow system

The answer is A, C, D, and E. Building commissioning, automatic occupancy lighting controls, daylight-responsive controls, and the use of LEDs are all within the interior designer's control for both design and specification.

A displacement ventilation system is a distribution system in which supply air originates at floor level and rises to return-air grilles in the ceiling. It must be designed by the mechanical engineer and is not part of the designer's work product. A variable refrigerant flow system must also be designed by the mechanical engineer.

56. An interior design project can obtain Leadership in Energy and Environmental Design (LEED) credits with the use of wood products that meet the criteria of

(A) FSC
(B) GBI
(C) Greenguard
(D) ISO 14000

The answer is A. The Forest Stewardship Council (FSC) is an international organization that oversees the development of national and regional standards based on basic forest management principles and criteria. It accredits certifying organizations that comply with its principles.

Green Building Initiative (GBI) is a not-for-profit organization dedicated to improving building performance and reducing climate impacts using their *Green Globes Assessment Protocol for Commercial Buildings*. It is not focused on wood use. Greenguard is a certification program of Underwriters Laboratories (UL) that develops tests for building materials, furniture, and furnishings (among other nonbuilding products) for emission standards to ensure they meet acceptable indoor air quality. ISO 14000 is a collection of standards and guidelines that cover issues such as performance, product standards, labeling, and other topics as they relate to the environment.

57. The methodology to evaluate the environmental impact of using a material or product in a building is known as

(A) a cradle-to-cradle review
(B) an impact assessment
(C) a life-cycle assessment
(D) a complete LEED study

The answer is C. A life-cycle assessment includes studying all aspects of a product's life cycle, including raw-material acquisition, manufacturing, use and maintenance, and disposal.

Cradle to Cradle (C2C) is a product certification program of McDonough Braungart Design Chemistry (MBDC) based on criteria of material health, material reutilization, renewable energy, water stewardship, and social fairness. Impact assessment is only part of a life-cycle assessment. There is no complete LEED study for materials or products.

58. The Green Label Plus designation is a program of the

(A) Building and Institutional Furniture Manufacturers Association
(B) Carpet and Rug Institute
(C) Resilient Floor Covering Institute
(D) Sustainable Forestry Initiative

The answer is B. The Carpet and Rug Institute has a voluntary program, called Green Label Plus, under which manufacturers have their carpets, cushions, and adhesives tested by an independent, certified laboratory for low chemical emissions. It also includes a chain of custody requirements and an annual audit of the testing laboratory. The other organizations do not have this designation.

59. A contractor has notified the interior designer that the building permit for the project has been denied. What course of action should the interior designer take first?

(A) notify the client of the situation and have the client contact their attorney
(B) instruct the contractor to work with the client to correct the objection
(C) with the client, file an appeal with AHJ's Board of Appeals
(D) determine what the objection is, make necessary changes, and have the contractor reapply

The answer is D. The first step should be to determine exactly why the permit was denied and, if possible, make necessary changes to the contract documents to satisfy the authority having jurisdiction (AHJ). Often, the objection is a minor thing and can easily be corrected. If not, an appeal to the Board of Appeals of the AHJ should be made.

60. What document allows the interior designer's client to occupy the project?

(A) AHJ Administrative Approval
(B) Certificate of Occupancy
(C) Final Inspection Report
(D) Temporary Certificate of Occupancy

The answer is B. A Certificate of Occupancy (sometimes called a Use and Occupancy letter) is required to be issued by the Authority Having Jurisdiction (AHJ) before the client can occupy the space. If there is a problem with one area of the space, a Temporary Certificate of Occupancy may be issued allowing the client to occupy just a portion as long as there is no danger. There is no such document as the AHJ Administrative Approval. A Final Inspection Report, if one is used by the AHJ, would not allow the client to occupy the space.

5 INTEGRATION WITH BUILDING SYSTEMS AND CONSTRUCTION

61. To increase the sound transmission coefficient (STC) of an interior partition, the interior designer should include which of the following? (Choose the four that apply.)

(A) sound absorbing material on one side
(B) double layer of gypsum wallboard
(C) sealant at all joints
(D) wood studs
(E) resilient channels
(F) insulation in the stud cavity

The answer is B, C, E, and F. A double layer of wallboard increases the mass of the partition, sealant stops sound leaks through cracks, resilient channels minimize vibration, and insulation dampens any sound waves within the partition. Placing sound-absorbing material on the "quiet" side of the partition may reduce noise in the quiet room through absorption of the sound waves, but it will not increase the actual STC value. Using wood studs or metal studs does not significantly affect the noise reduction through the partition.

62. The MOST economical way to create a column-free space in a new home addition is to use

(A) plywood web I-joists
(B) double joists
(C) laminated veneer lumber
(D) glued laminated lumber

The answer is A. Plywood web I-joists are specifically designed to replace joists to allow for longer spans economically because they use relatively thin plywood for the webs and small solid wood pieces for the flanges in an efficient configuration.

Double joists would not be economical or practical to increase span distances because twice as much lumber would be required, and locating long joists is difficult. Laminated veneer lumber is commonly used as a replacement for headers, studs, and shorter beams, not as a replacement for joists. Glued laminated lumber would not be appropriate because it is used for long-span beams that support other structural members.

63. Typically, the MOST effective action an interior designer can take to maximize daylighting is to

(A) change the type of glazing
(B) increase the head height of glazing
(C) install a light shelf
(D) specify a high-reflectance paint

The answer is D. The interior designer can typically influence only the surface finishes and reflectance of rooms; therefore, to maximize daylighting, the interior designer should specify a high-reflectance paint. Light shelves and glazing are usually part of the building's architecture and are outside the interior designer's control.

64. An interior designer is planning a laboratory space adjacent to a restroom. What type of partition construction would best accommodate the plumbing and other services necessary to serve both spaces?

(A) metal studs 6 in (152.4 mm) deep with gypsum board on both sides
(B) double masonry with space between
(C) demountable partition
(D) a chase wall

The answer is D. A chase wall consists of two runs of studs separated by a distance required to allow for installation of all the services. To accommodate the plumbing pipes, gas piping, electrical conductors, and other services typically required for a laboratory and restroom, a chase wall is the appropriate choice.

Metal studs 6 in (152.4 mm) deep would not provide sufficient space for the services and would make horizontal runs of piping difficult. Double masonry could provide the necessary space but is expensive to construct. A demountable partition is only appropriate for spaces that need to be reconfigured frequently, and it would not provide space for services.

65. In the partial section of a glass block partition shown in the illustration, drag the material labels from the left into the boxes on the section detail. Some material labels may not be used.

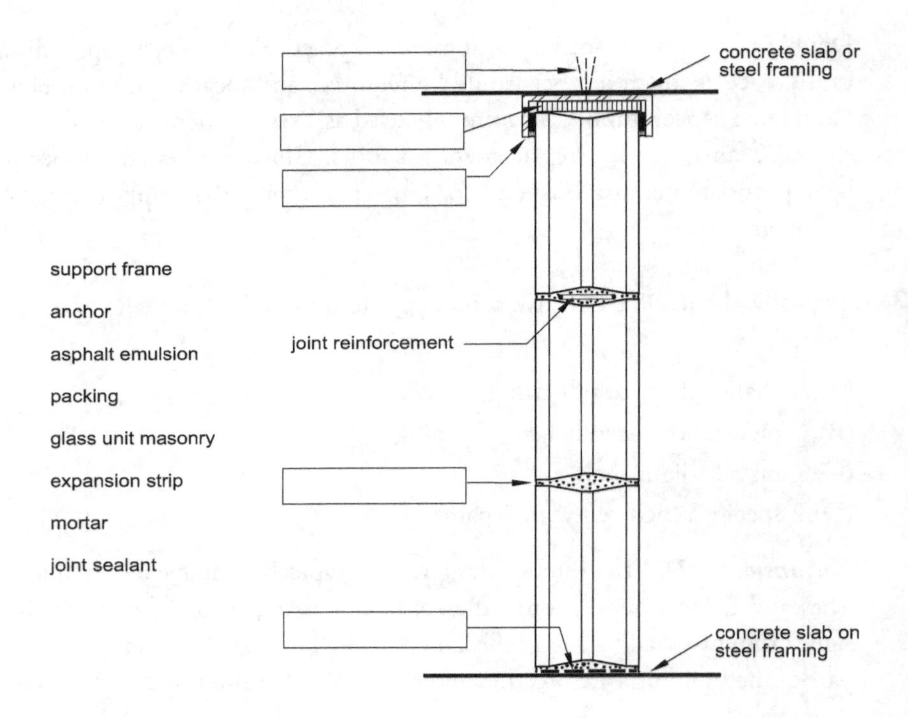

The answer is as follows. The section shows a typical interior glass block (glass unit masonry) partition. The first course of block is laid with mortar over an asphalt emulsion. Subsequent courses are laid with mortar, and horizontal joint reinforcement is provided at every other horizontal joint, at a minimum of 16 in (406 mm) on center. At the top course below the structure (either concrete or steel framing), an expansion strip is used to accommodate structural deflection as well as any expansion that may occur in the wall itself. An anchor is used to attach the support frame to the structure. Packing is placed between the block and the frame and finished with joint sealant.

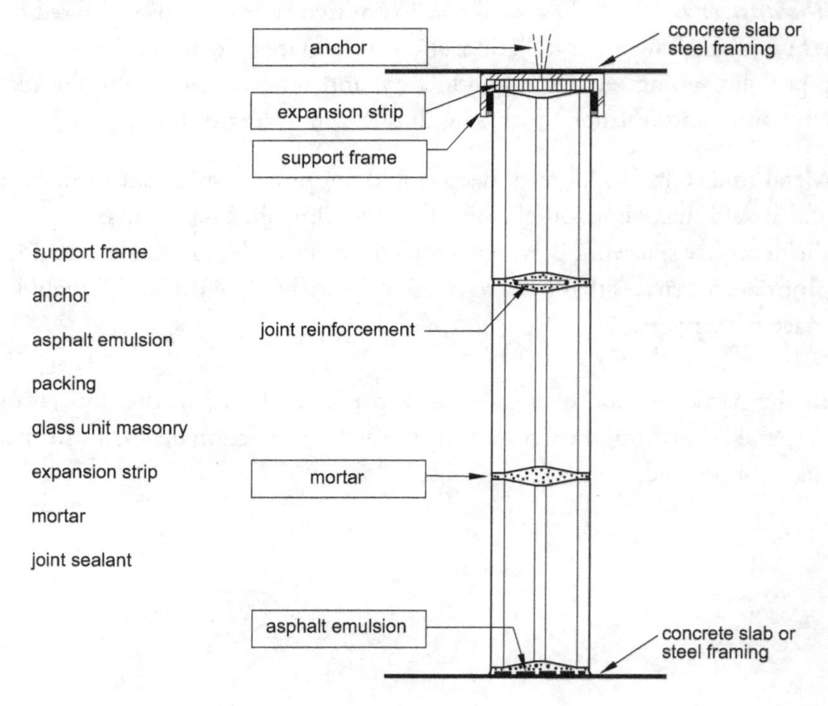

66. In a small lecture hall, it would be BEST to avoid

(A) a sound amplification system
(B) a vaulted ceiling
(C) carpet
(D) parallel walls

The answer is B. A vaulted ceiling would focus sound reflections into one concentrated area and produce annoying echoes or quiet spots in the hall.

67. What is the unit used to measure luminance?

(A) footcandles
(B) footlamberts
(C) candelas
(D) lumens

The answer is B. Luminance is the brightness reflected or transmitted from a projected area (i.e., only the area that you see when looking at the source). In other words, it is the brightness of a direct glare source. In the older inch-pound system, luminance is measured in footlamberts (ftL), where one footlambert is equal to the reciprocal of π in candelas per square foot (written $1\ \text{ftL} = 1/\pi\ \text{cd/ft}^2$).

68. When working on a home remodeling project, which type of electrical conductor is the interior designer MOST likely to encounter?

(A) armored cable (AC)
(B) electrical metallic tube conduit (EMT)
(C) intermediate metal conduit (IMC)
(D) nonmetallic sheathed cable (NM)

The answer is D. For residential construction, NM, or Romex®, is typically used. This consists of two or more plastic-insulated conductors and ground wire surrounded by a moisture-resistant plastic covering. It is relatively inexpensive and easy to install through wood-frame construction. The other types of conductors are used in commercial construction.

69. One of the MOST frequent lighting problems in traditional drafting rooms is

(A) veiling reflection
(B) direct glare
(C) visual comfort
(D) excessive brightness ratio

The answer is A. Any lights in the ceiling of a drafting room are reflected off plastic triangles, parallel bars, and computer screens, causing veiling reflections.

70. What term is used to denote the member supporting the treads in a wood stairway?

(A) carriage
(B) header
(C) saddle
(D) stringer

The answer is A. As shown in the following illustration, the carriage supports the treads in a wood stairway.

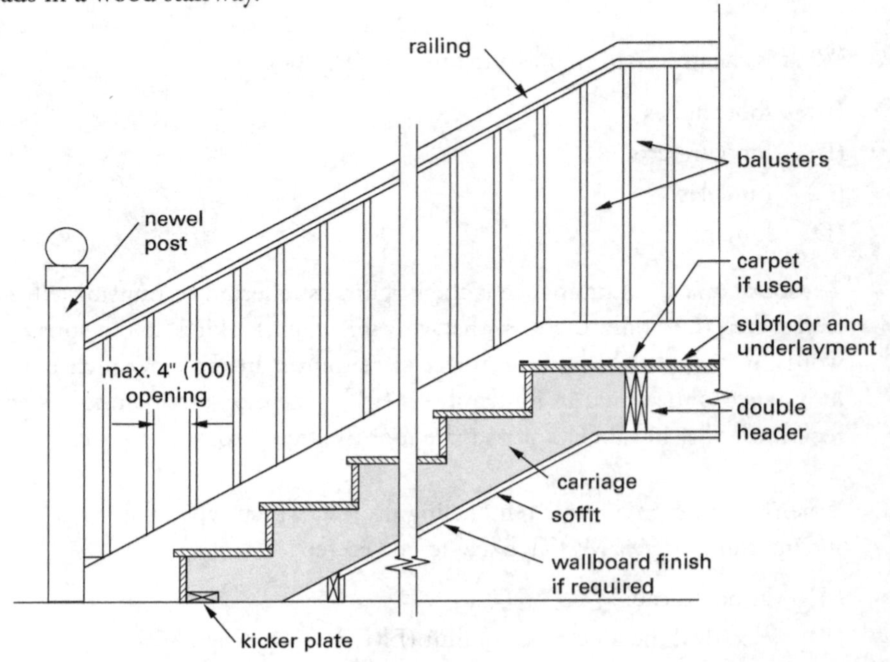

71. What would be the BEST combination of dimensions for a stair serving a sleeping loft in a condominium where space is at a premium and the floor-to-floor dimension is 9 ft 4 in (2845 mm)?

(A) 7 in (178 mm) riser, 10 in (254 mm) tread
(B) 7 in (178 mm) riser, 11 in (279 mm) tread
(C) 8 in (203 mm) riser, 10 in (254 mm) tread
(D) 8 in (203 mm) riser, 11 in (279 mm) tread

The answer is A. If space is at a premium and the floor-to-floor dimension is fixed, minimize the number of treads and the width of each one so as to achieve the shortest possible total run. Options B and C with the 8 in (203 mm) riser cannot be used because the code maximum for residential uses is 7$^3/_4$ in (197 mm). The 7 in (178 mm) riser will work, so select the one with the narrowest tread, 10 in (254 mm). A 10 in (254 mm) tread is acceptable by code in residential occupancies.

72. A compact filing system is an example of a

(A) dead load
(B) dynamic load
(C) lateral load
(D) live load

The answer is D. Live loads include the loads of people, furniture, snow, and equipment.

73. A slip joint at the top of a partition is required to account for

(A) clearance
(B) movement
(C) tolerance
(D) structure

The answer is B. A slip joint allows the floor above a partition to move without damaging the partition.

74. A programming report for a clinic in a stand-alone building states that controlling noise transmission is of critical importance. In designing the space plan, the interior designer can MOST effectively control noise by

(A) locating noise-sensitive spaces away from the mechanical room
(B) providing corridors between as many rooms as possible
(C) putting buffer rooms between noisy spaces and quiet spaces
(D) separating noisy and quiet spaces with concrete block partitions

The answer is C. Because there may be several noisy spaces in a clinic, the best strategy in basic planning is to separate the noisy and quiet spaces with buffer rooms such as storage closets. Later, in detailing, sound control partitions can be provided where sound control is critical.

The mechanical room is unlikely to be the only noisy space in a clinic, so locating noise-sensitive spaces away from the mechanical room will probably not solve the problem. Providing as many corridors as possible between rooms may help, but this is an inefficient use of space. Using a sound control partition is a good way to limit sound transmission, but the question asks about space planning, not detailing or material selection.

75. What is included in the rise of a stair?

(A) vertical distance from one nosing to the next
(B) average height of a step
(C) distance from finish floor slab to finish floor slab
(D) number of steps between landings

The answer is C. Refer to the following illustration. Option A describes a riser.

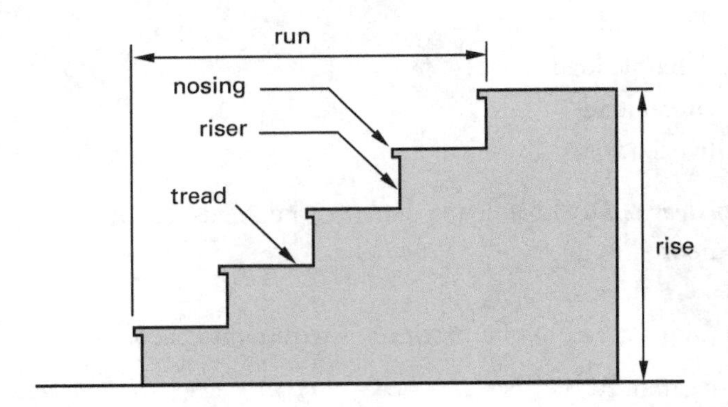

❻ INTEGRATION OF FURNITURE, FIXTURES, AND EQUIPMENT

76. When the interior designer is ordering furniture for a project, the purchase orders are sent

(A) before the sales agreement
(B) after the client decides on the purchases
(C) after confirmation of the order but before shipping
(D) after the sales agreement and before the acknowledgment

The answer is D. After receiving the purchase order, the manufacturer sends an acknowledgment or confirmation to the interior designer. The client must first sign a sales agreement or contract obligating the client to pay for the furniture. Then the interior designer fills out one or more purchase orders for the merchandise and sends the documentation to the manufacturer.

77. In what order of importance should tables for a college library be selected?

(A) flammability, design, then comfort
(B) finish, flammability, then design
(C) durability, cost, then design
(D) quality, comfort, then finish

The answer is C. Because institutional furniture takes much abuse and must last a long time, its durability is important. This suggests option C or D because quality can be considered a measure of durability. However, cost is usually an important factor in furniture selection for this type of client, so option C is the best.

78. When completing construction drawings for a large brokerage firm, the interior designer must communicate the requirements of a large number of pieces of office equipment. What is the best location for this type of information?

(A) floor plans
(B) schedules
(C) specifications
(D) shop drawings

The answer is B. Information on office equipment would include a variety of data types such as name, manufacturer, size, weight, and electrical requirements. A schedule of equipment would be the best way to convey the information required.

79. What are two characteristics of acoustic panels necessary to make them effective?

(A) They must have at least 1 in (25 mm) of sound-absorbing material and use permeable fabric.
(B) The fabric must be hydrophobic, and the sound-absorbing material must be at least 3/4 in (19 mm) thick.
(C) The fabric should be unbacked and made of hydrophilic material.
(D) They should use tackable acoustic fiberglass as a filler and have hydrophilic fabric.

The answer is A. Acoustic panels should be at least 1 in (25 mm) thick and have permeable fabric to allow sound waves to pass through. As such, the panels should also not be backed. Hydrophobic fabric should be used as it doesn't absorb and hold moisture that could cause sagging and distortion. Hydrophilic fabrics absorb and retain moisture and should be avoided. While tackable acoustic fiberglass can be used, it should not be covered with a hydrophilic fabric.

80. What is the primary advantage of systems furniture?

(A) built-in ambient and task lighting
(B) interchangeable panels with work surfaces
(C) flexibility and efficient space use
(D) integrated wire management and power outlets

The answer is C. As a collection of component parts, including dividing panels, systems furniture makes efficient use of space compared with private offices or freestanding furniture and provides flexibility in assembling what pieces are required for a worker and how individual stations are laid out.

Ambient and task lighting may be important parts of systems furniture but not its primary advantage. Panels and work surfaces are only two parts of system furniture and are not advantages in themselves. As with lighting, wire management and power outlets are important but not the primary advantage of systems furniture.

81. Which type of partition system offers the most flexible and easy-to-configure system to create individual rooms and spaces?

(A) modular wall systems
(B) nonprogressive demountable partitions
(C) progressive demountable partitions
(D) operable partitions

The answer is A. A modular wall system is a series of components used to quickly and easily subdivide a space. The exact details of a modular wall system vary slightly depending on the manufacturer. Some have a framework of aluminum or steel studs on which panels are hung and others use gypsum wallboard or embossed steel over polystyrene panels, which are connected to a post system. Other finishes include vinyl wrapped gypsum, laminate, tackable panels, and glass. Some are stabilized with vertical posts and others are supported with C-channels attached to the ceiling trim. Accessories include door and window openings. Many modular wall systems can be configured by maintenance personnel without the need for a contractor. They differ from demountable partitions in that they are easier to install and generally do not appear to be a standard fixed partition.

Demountable partitions systems are also composed of individual parts but are generally attached more firmly to the building substrate. With a nonprogressive demountable system, the panels are independent and can be removed or replaced individually. A progressive system requires that the first panel be installed before the second, and so on. An operable partition is basically a large special door fixed along a track that can be opened and closed frequently.

82. Customer's own material (COM) refers to

(A) material that is custom ordered from a furniture manufacturer for a selected item

(B) fabric not part of a manufacturer's standard fabric line that is specified separately by the designer

(C) furniture that will be supplied by the customer for installation by the furniture installer

(D) fabric that the designer's client has on hand that they want to use for a chair or sofa

The answer is B. Customer's own material (COM) is fabric ordered separately from a source other than a furniture manufacturer whose standard product line may not have the type or color of fabric that the designer or client prefers. The fabric is ordered from another source and shipped to the furniture manufacturer to be used for an upholstered item, usually a chair or sofa. COM is not custom ordered from the furniture manufacturer; any fabric as part of a manufacturer's standard product line is essentially a custom order. COM is fabric, not furniture, and the client does not supply fabric for use on the project.

83. When shipping furniture with a common carrier, "FOB destination" means

(A) The title to the furniture is transferred at the factory and the transportation cost is paid by the buyer.

(B) The client is responsible for arranging for the shipment of furniture to its final destination and for carrying shipping insurance.

(C) The interior designer hires a common carrier to ship furniture to a local dealership for installation.

(D) The seller (manufacturer) is responsible for shipping and recovering any damage or loss during shipment.

The answer is D. FOB means *free on board,* and "FOB destination" means that the manufacturer pays for loading the goods on the truck or train and is also responsible for shipping and recovering any damage or loss during shipment. This is also known as a designation contract.

If the title is transferred at the factory, it is known as "FOB factory." The manufacturer is not responsible for any loss or damage during shipping, and the buyer should carry insurance for the goods while in transit. The client is not responsible for arranging furniture shipment; that is usually the responsibility of the FF&E vendor or furniture dealer. Likewise, the interior designer does not hire shipping companies.

84. Which of the following are important maintenance issues to consider when selecting fabric for a commercial interior? (Choose the four that apply.)

(A) repairability
(B) flammability
(C) crocking resistance
(D) cleanability
(E) abrasion resistance
(F) chemical resistance

The answer is A, C, D, and E. Repairability is the ability of a fabric to be repaired or replaced when damaged. For example, it is easier to replace a fabric rather than buying an entirely new piece of furniture for maintenance. Crocking resistance is the resistance of a colored textile to transfer its color from its surface to other surfaces. Low crocking resistance means the fabric would have to be replaced more often. Cleanability refers to the ease with which a fabric can be cleaned using whatever methods are appropriate. It is one of the most important factors in maintaining furniture. Abrasion resistance is important so a fabric maintains its original appearance after repeated use.

Flammability is a safety issue and not a maintenance issue. Chemical resistance is a material's resistance to damage or change of finish resulting from exposure to chemicals. For fabric, stain resistance is a more appropriate issue to consider.

85. The main advantage of a client supplying their own freestanding equipment instead of having the contractor purchase it is that the

(A) client is assured of getting exactly what is wanted
(B) interior designer is relieved of the responsibility for specifying
(C) client avoids the markup cost that a contractor would apply
(D) contractor does not have to spend time looking for the right equipment

The answer is C. The main advantage is that when the client supplies the equipment, they purchase it directly, so the contractor does not automatically add money for overhead and profit as they would for other construction supplies and labor. While the other options may be advantages, the cost advantage is the most important one.

7 CONTRACT ADMINISTRATION

86. During construction, a client requests that a door be added to a partition. Given that the client agrees to the cost of adding the door, the interior designer administering the associated contract should issue a(n)

(A) addendum
(B) construction change directive
(C) change order
(D) work directive

The answer is C. A change order (signed by the contractor, client, and interior designer) would be issued for any change to a project's contract during construction that requires additional time or money. A construction change directive is used to order a change in the work when there may be a change in cost or time, but the client and contractor cannot yet agree on the price. Once a price is set, the interior designer would then issue a change order. The interior designer would use an addendum to make changes before the project's contract is signed. There is no such document as a work directive.

87. Owners set bidder requirements in the

(A) advertisement to bid
(B) invitation to bid
(C) prequalification of bidders
(D) prebid conference

The answer is C. In the prequalification of bidders, owners can set restrictions on such factors as the bidders' years of experience, bonding capacities, firm's size, and so on.

88. During project closeout, the interior designer is required to do which of the following? (Choose the three that apply.)

(A) develop a punch list
(B) collect and send all warranties to the owner
(C) determine the date of substantial completion
(D) provide the owner with bonds
(E) finalize the placement of the owner's furniture
(F) send all pertinent maintenance contracts to the owner

The answer is A, C, and E. The interior designer is responsible for developing a punch list, determining the date of substantial completion, and finalizing the placement of the owner's furniture. Submitting all warranties, operating instructions, bonds, and maintenance contracts to the owner is the responsibility of the contractor, not the interior designer.

89. Six weeks after construction has started on a large restaurant, the contractor installs a service counter that the owner thinks is built incorrectly. After everyone involved has reviewed the contract documents, who is responsible for making the determination about the counter?

(A) contractor
(B) interior designer
(C) owner
(D) interior designer and owner

The answer is B. The interior designer is the sole interpreter of the contract documents' requirements, as long as decisions can reasonably be based on the content of the drawings, specifications, shop drawings, and other contract documents.

90. At the end of a project, the project manager's primary responsibility is to

(A) monitor quality
(B) coordinate payment
(C) direct closeout
(D) hold and document meetings

The answer is C. The closeout may require the project manager to monitor quality, coordinate final payment to contractors and vendors, and hold and document meetings. However, the project manager's primary responsibility at the end of a project is to direct closeout so that *all* the closeout activities, not just these individual activities, are completed.

91. What should the interior designer do FIRST if a client decides to make major revisions after a project has been tendered but before construction has started?

(A) Tell the contractor not to proceed until the issues have been resolved.
(B) Return any shop drawings to the contractor, and tell the contractor that revisions will be forthcoming.
(C) Advise the client that making major revisions may delay the job and increase its cost and that additional fees will be charged for design and drawing revisions.
(D) Estimate the amount of time and extra fees needed to make the revisions, and suggest that the client reconsider making major changes.

The answer is C. This question requires knowledge of the word *tendered* as well as of the procedures for handling these types of changes. *Tender* is a term used in England and often in Canada that means the same as *to bid.* Making changes after a project has been bid can be a major problem, and the interior designer should be sure the client understands the implications of time delays and cost changes.

92. An interior designer has reviewed and approved the shop drawings for the installation of a door with glazing between two existing block walls. During construction, the subcontractor responsible for installing the door discovers that it will not fit as planned. Who is responsible for correcting and paying for the error?

(A) general contractor
(B) doorframe fabricator
(C) interior designer
(D) subcontractor

The answer is A. Although the interior designer reviewed the shop drawings, the general contractor is responsible for taking field measurements for new construction, and is also ultimately responsible for the accuracy of the drawings. The interior designer's review is only to verify general compliance with design intent. Therefore, the general contractor would be responsible for correcting and paying for the error.

93. During a construction site visit, the interior designer notices that a finish subcontractor is not installing a ceiling material properly. The interior designer's first response should be to

(A) inform the owner in writing of the situation
(B) withhold the appropriate amount on the contractor's application for payment
(C) tell the contractor the work is not in conformance with the contract documents
(D) notify the subcontractor that the material is not being properly installed

The answer is C. According to the contract documents, the interior designer's communication with subcontractors and material suppliers must be through the contractor.

94. After the interior designer gets a notarized application for payment from the contractor, what must the designer do before sending the application to the client for payment?

(A) Request copies of invoices received from subcontractors and material suppliers and review them to see if the costs coincide with the contractor's costs on the application for payment. If so, then certify the application and send it to the client.
(B) Visit the jobsite for a careful examination of the work and stored materials to make absolutely sure the contractor's application represents the actual condition of the project.
(C) Verify that the application for payment has been submitted at least 10 days before the date established for each payment in the *General Conditions of the Contract for Construction*, make a site visit, review the money requested, and certify the application.
(D) Review the application for payment and certify that the work has progressed to the point indicated on the application and that, to the best of the designer's knowledge, the quality of the work meets the requirements of the contract documents and no extra money has been requested for work not done or materials not in storage.

The answer is D. The designer must certify the application, which constitutes an acknowledgment by the interior designer that the work has progressed to the point indicated and that, to the best of the interior designer's knowledge, information, and belief, the quality of the work is according to the contract documents. Certification is not a representation that the designer has made exhaustive on-site inspections or that the designer has reviewed construction methods, techniques, or procedures.

The designer is not obligated to review subcontractors' and material suppliers' invoices or requisitions, nor is the designer obligated to make exhaustive on-site inspections. The application for payment must be submitted 10 days before the date of payment but does not necessarily require the designer to make a separate site visit.

95. Who is responsible for organizing and holding project meetings?

(A) client
(B) contractor
(C) project manager
(D) office assistant

The answer is C. The project manager should organize and conduct regular project meetings to be attended by the contractor, client, and other parties that are involved in critical work at the time of the meeting, such as the mechanical engineer, lighting designer, or electrical engineer. The office assistant may help in taking notes. These are primarily design and progress meetings. The contractor may hold separate meetings that involve the project superintendent, construction workers, and material suppliers.

96. What is the interior designer's purpose when performing construction observation? (Choose the four that apply.)

(A) keep the owner informed
(B) determine if the contractor is following adequate safety procedures on the jobsite
(C) endeavor to guard the owner against defects in the work
(D) determine, in general, if the work is progressing in such a way that, when completed, will be in accordance with the contract documents
(E) verify that the materials stored on the jobsite and elsewhere matches the itemizations on the applications for payment
(F) become generally familiar with the progress and quality of the work

The answer is A, C, D, and F. The purpose of the designer's construction observation is to become generally familiar with the progress and quality of the work and to keep the owner informed, to endeavor to guard the owner against defects and deficiencies in the work, and to determine, in general, if the work is progressing in such a way that, when completed, it will be in accordance with the contract documents.

The interior designer is not required to make exhaustive inspections of the jobsite. The number and timing of visits are left to the judgment of the designer. The interior designer is not responsible for safety on the jobsite or for the safety means and methods followed by the contractor. The interior designer is also not required to verify quantities of stored materials.

97. In consultation with a lighting designer, an interior designer has designed a series of custom suspended light fixtures. The vendor who is manufacturing the fixtures has sent the shop drawings to the general contractor for review and distribution to the appropriate parties. In what sequence should the shop drawings be reviewed and passed on?

(A) interior designer, electrical engineer, general contractor
(B) lighting designer, interior designer, general contractor
(C) electrical engineer, lighting designer, interior designer, general contractor
(D) interior designer, lighting designer, interior designer, general contractor

The answer is D. When reviewing submittals, the general contractor first reviews and approves them. By doing this, the general contractor represents that field measurements have been verified, materials have been checked and other construction criteria have been coordinated. The drawings are then sent to the interior designer to see if they are in conformance with information given to the vendor and to see if they follow the design intent. The interior designer's review does not relieve the contractor of any responsibilities. In this situation, the interior designer would then forward the shop drawings to the lighting designer, who would review them and then send the drawings back to the interior designer with any comments or corrections. Finally, the interior designer would send them back to the contractor marked up as "no exceptions taken," "make corrections noted," or "revise and resubmit" as appropriate. The contractor then sends them back to the vendor. In this case, the electrical engineer would not be involved.

98. Who is responsible for the disposal of a mock-up that cannot be incorporated into the work after it has been reviewed and approved?

(A) client
(B) interior designer
(C) contractor
(D) individual stated in the specifications

The answer is D. A mock-up is a full-sized sample of a portion of the construction, commonly built on the jobsite. It can be either separate from the building or, if approved, can be integrated into the building. Normally, a mock-up is called for in the specifications and paid for by the client. The specifications state what kind of mock-up is required, whether one of inexpensive, temporary materials or one like the finished construction that can be incorporated into the work. The specifications should also state who is responsible for their disposal. If the mock-up is large and temporary or not approved, the specifications should probably state the contractor should dispose of it as the contractor has access to trash facilities on the jobsite or where it is built. If the mock-up is small, the client or interior designer may want to keep it.

99. When does the interior designer first inspect the contractor's completed work and make a punch list/deficiency list?

(A) after the contractor has notified the interior designer the work is complete
(B) when the interior designer issues the final certificate for payment
(C) before the contractor has made their inspection of work to be completed
(D) during a walk-through with the contractor to determine uncompleted work items

The answer is A. The contractor begins project closeout procedures by notifying the interior designer in writing that work is complete and submits a comprehensive list of items still to be completed or corrected. This is the procedure detailed in AIA Document A201, *General Conditions of the Contract for Construction* if that is the document being used (with the words Interior Designer substituted for the word Architect). The interior designer then makes an inspection to determine whether the work is substantially complete and whether there are more items that need to be added to the list. This is called the punch list (or deficiency list in Canada). The contractor must correct these items, after which another inspection by the interior designer is called for. If the final inspection shows that the work is substantially complete, the interior designer issues a certificate of substantial completion. When the certificate of substantial completion is issued, the final application for payment is processed, and the interior designer issues a final certificate for payment.

100. How can the interior designer control the amount of time that the designer has to review submittals such as shop drawings, samples, and product data?

(A) The contractor determines the amount of time available for submittals, and the interior designer must abide by these times.

(B) The interior designer can take whatever time is required to review any given submittal.

(C) The interior designer should include in the specifications the procedure for making submittals, including the time the contractor must allow for the designer's review.

(D) The contractor must request what time the designer needs for each submittal at the beginning of the project.

The answer is C. Although the *General Conditions of the Contract for Construction* and the Owner-Designer agreement state that the interior designer must review submittals with reasonable promptness, neither of them states a specific time limit. However, the *General Conditions of the Contract for Construction* does require that the contractor prepare a construction schedule that must include a schedule of submittals, which must allow the interior designer reasonable time to review the submittals. The interior designer should include in Division 01 of the specifications the procedure for making submittals, including the time the contractor must allow for the interior designer to review submittals. The contractor will then include this time as part of the construction schedule. The interior designer must conform to the time in the contractor's schedule but can modify it by using this procedure.

The interior designer doesn't necessarily need to abide by the contractor's times if more or less time is included in Division 01 by the designer. The designer cannot take whatever time they feel like; it must be determined in the contractor's schedule. The contractor does not request the time from the designer; this only comes from being included in Division 01 of the specifications.

MOCK EXAM

Do not use reference books while taking this mock exam. Besides this book, you should have only pencils, scratch paper, and a calculator. (For the actual exam, these will be provided and should not be brought into the site.)

Exam time limit: 4.0 hours

❶ PROJECT ASSESSMENT AND SUSTAINABILITY

1. An interior designer has determined the required gross occupant area required for a client planning to lease space in an office building. Using the Building Owners and Managers Association (BOMA) standard ANSI/BOMA Z65.1, *Office Buildings: Standard Methods of Measurement*, which of the following would be used to determine the amount of lease space the client will be paying for?

(A) building efficiency ratio
(B) building load factor
(C) R/U ratio
(D) rentable area

2. During design development of a commercial office space, the interior designer can minimize energy use by coordinating space planning with

(A) daylighting zones
(B) HVAC zones
(C) occupant sensor controls
(D) time-switch controls

3. In most situations, the application for a building permit is made by the

(A) building owner
(B) client
(C) contractor
(D) interior designer

4. Which of the following constitutes due diligence in the field of interior design?

(A) thoroughly understanding and documenting the space in which a client's project will be located and its surrounding context
(B) reviewing the legal and regulatory components of a new project before design begins
(C) using qualified structural, engineering, and fire protection consultants to investigate an existing space before design begins
(D) reviewing and verifying site conditions that the client has reported to the interior designer

5. Which of the following facts about the surrounding neighborhood are relevant to whether a building can achieve the LEED for Commercial Interiors (LEED-CI) certification? (Choose the four that apply.)

(A) All surrounding streets are collector or arterial.
(B) The building is in a high-density, walkable community.
(C) The number of guaranteed tenant parking spaces is the minimum allowed by zoning requirements.
(D) The site has sufficient access to sunlight from the south.
(E) The site has sufficient bicycle facilities.
(F) Public transportation is located within ¼ mi (0.4 km) of the project site.

6. Which are the MOST important factors in determining the space required for a nurses' station in a hospital?

(A) number of patient contacts and files
(B) work surface and storage requirements
(C) number of moveable carts and space requirements for writing reports
(D) electrical and communication equipment requirements

7. In investigating the existing conditions of an older building before beginning space planning, a project manager would probably need to seek expert consulting assistance to determine the

(A) number of supply air diffusers in the space
(B) existence of floor-mounted electrical outlets in an open space
(C) presence of adequate water pressure to add a sink in a washroom
(D) feasibility of opening a double-wide doorway in an existing loadbearing wall

8. A chef is considering the purchase of an abandoned church, with the intention of turning it into a restaurant. The property's zoning designation allows this type of use. The chef has obtained a preliminary site plan and permission to access the site from the current owner and hires an interior designer to assist with a feasibility study to examine the possibility of converting the old building to the new use. Which of the following should the interior designer do first?

(A) Check local zoning ordinances and analyze the site to determine if there is enough space available for the parking required.
(B) Assist the restaurateur in developing a program and check the space requirements against the area and layout of the existing structure.
(C) Research the history of the church.
(D) Complete a code review of the existing building and develop a preliminary plan for renovation.

9. An office space is being measured according to ANSI/BOMA Z65.1, *Office Buildings: Standard Methods of Measurement*. The designer should measure to which of the following when the curtain wall glazing covers more 70% of the exterior wall and a convector cover is 12 in (305 mm) high?

(A) face of the convector cover on the exterior wall
(B) inside face of the glazing of the exterior wall
(C) center of the curtain wall
(D) outside face of the glazing of the exterior wall

10. According to ANSI/BOMA Z65.1, *Office Buildings: Standard Methods of Measurement*, which of the following comprise the building common area? (Choose the four that apply.)

(A) private tenant stairways
(B) elevator lobbies
(C) tenant office space
(D) mechanical rooms
(E) public corridors
(F) vending areas

11. An interior designer is working for a client who has leased space with a gross occupied area of 4500 ft^2 (418 m^2). The leasing agent tells the interior designer that the load factor for the building is 1.15. To the nearest square foot (square meter), the amount of space that the client will pay rent on is ________ ft^2 (________ m^2). (Fill in the blank.)

12. What are the best actions an interior designer can take to assist a client in developing plans to minimize water use for a new restaurant? (Choose the four that apply.)

(A) suggest the mechanical engineer develop a system for graywater use
(B) specify only restroom fixtures that are approved by WaterSense
(C) verify with the kitchen consultant that high-efficiency pre-rinse spray valves are used
(D) suggest that the temperature setting for dishwashing be reduced to 130°F (54°C)
(E) suggest the client develop an ongoing program to find water leaks
(F) make sure dishwashers and ice machines are ENERGY STAR qualified

13. What are the most common sources of poor indoor air quality? (Choose the four that apply.)

(A) contaminants from outdoor sources
(B) biological contaminants
(C) poor building maintenance
(D) poor ventilation
(E) indoor chemical contaminants
(F) sick-building syndrome

14. An interior designer is planning an office space that will include a room with computer equipment and a moderate number of workstations. The computer room will use an extensive raised flooring system to service the computer equipment and as a result will have a lower than normal finished ceiling height. What is the best measure the designer can recommend to improve energy efficiency?

(A) suggest the mechanical engineer use an all-air type system to cool the equipment
(B) require that the HVAC system conform to the *International Energy Conservation Code*
(C) space plan so the computer equipment area is in the interior of the building
(D) suggest the mechanical engineer use a displacement ventilation system

15. Which of the following indoor contaminants would require the assistance of a licensed professional contractor to mitigate?

(A) formaldehyde
(B) commercial cleaning compounds
(C) vermiculite
(D) volatile organic compounds

16. An interior designer has been retained by the developer of a suburban retail complex to design the interior of a mid-size department store. The interior designer is working with the architect of the complex along with the architect's consulting engineers. Which project participants are likely to have the most influence on the designer's creative work? (Choose the three that apply.)

(A) leasing agent
(B) developer
(C) department store owner
(D) public
(E) store employees
(F) architect

17. Which of the following flooring material choices is a good renewable option?

(A) linoleum
(B) carpet made with PET
(C) rubber
(D) vinyl

18. An interior designer is assisting a law firm in selecting one of two buildings in which to lease space. After programming, the designer determines that the client will need 6500 ft^2 (604 m^2) of net assignable area. In building A, the designer estimates that the efficiency factor will be approximately 80%. In building B, because of the irregular shape of the floor plate, the designer estimates that the efficiency factor will probably be closer to 75%. The leasing agent in building A tells the designer that the building load factor is 1.23, while the leasing agent in building B tells the designer that the building load factor is 1.20. Of the two buildings, to the nearest 100 ft^2 (100 m^2), the smaller rentable area is __________ ft^2 (____________ m^2). (Fill in the blank.)

19. What strategies can the interior designer recommend to a client to maintain good indoor air quality after occupancy? (Choose the four that apply.)

(A) select materials with low VOCs
(B) suggest a no-smoking policy for the space
(C) plan a separate room for activities with high concentrations of pollutants
(D) specify independent building commissioning
(E) have the client monitor individual spaces for contaminants
(F) verify that minimum outdoor air ventilation is being provided

20. In reviewing the floor plan of an existing space to determine which corridors, partitions, and doors can be reused, what is the designer most likely to overlook?

(A) floor level changes
(B) protruding objects
(C) ramps
(D) maneuvering space at doors

21. If an interior designer thinks that a project is in a high seismic activity area, what course of action should the designer first take concerning ceiling systems?

(A) ask the local authority having jurisdiction what category the building is in
(B) consult with a structural engineer to verify what seismic category the project is in
(C) include ceiling details on the construction drawings that comply with the IBC
(D) specify that the ceiling installation contractor provide the required seismic details

22. An interior designer is preparing a cost evaluation for a project during the programming stage. The opinion is based upon a previous similar project. Which of the following is a known factor that will add cost to the project?

(A) a contingency
(B) a premium
(C) an additive alternate
(D) an upcharge

23. Early in the programming process, the goals and objectives of the client would include which of the following?

(A) how the client wants to organize the project
(B) which of the groups using the project are of primary importance
(C) what the client wants to achieve and why
(D) how many spaces are needed and what their areas must be

24. A change in use in an existing building requires the interior designer to first research

(A) the allowable square footage established by the building code for the new use
(B) the number, location, and separation of the exits to the outside
(C) maximum travel distance of the old use compared with the new use
(D) the existing construction type

25. Which of the following can the interior designer implement to support energy conservation in a library project? (Choose the four that apply.)

(A) do not allow mechanical engineers to use any systems that include CFC-based refrigerants
(B) specify automatic occupancy lighting controls
(C) plan reading rooms and study areas below north-facing skylights
(D) design automated shading systems in window openings
(E) place book stacks near south facing windows to moderate glare
(F) use displacement ventilation throughout the building

26. An interior designer is assisting a client in selecting an existing building location among many in which to locate a small retail development. The client has indicated that the project should receive a Leadership in Energy and Environmental Design (LEED) rating. Which of the following should the designer do to help achieve this goal?

(A) look for buildings with insulating glass
(B) review the existing partition layout to verify possible reuse
(C) verify that parking amounts to less than 50% of zoning requirements
(D) select a location that has been designated a brownfield site

❷ PROJECT PROCESS, ROLES, AND COORDINATION

27. With whom should the interior designer consult to determine where to locate freestanding stacks in a library?

(A) mechanical engineer
(B) library planner
(C) structural engineer
(D) fire protection designer

28. In specifying new window coverings and tracks to replace building-standard window coverings, what coordination is MOST important for the interior designer to undertake? (Choose the four that apply.)

(A) verifying that the building owner approves of the change in the building's exterior appearance
(B) asking the electrical engineer if the new light reflectance is detrimental
(C) determining if the mechanical engineer objects to the replacement plans
(D) getting approval from the client concerning the color of the coverings
(E) verifying with the manufacturer that flammability does not pose a problem
(F) checking with the architect to see that the new coverings do not adversely affect the heating of the glass

29. In designing a new space, the interior designer can influence life safety the MOST by coordinating with consultants concerning

(A) compartmentation
(B) fire detection
(C) smoke control
(D) sprinklers

30. Who is responsible for showing the telephone and data conduit runs on the drawings for a commercial design project?

(A) architect
(B) electrical engineer
(C) interior designer
(D) telephone installer

Questions 31 and 32 refer to the following drawing.

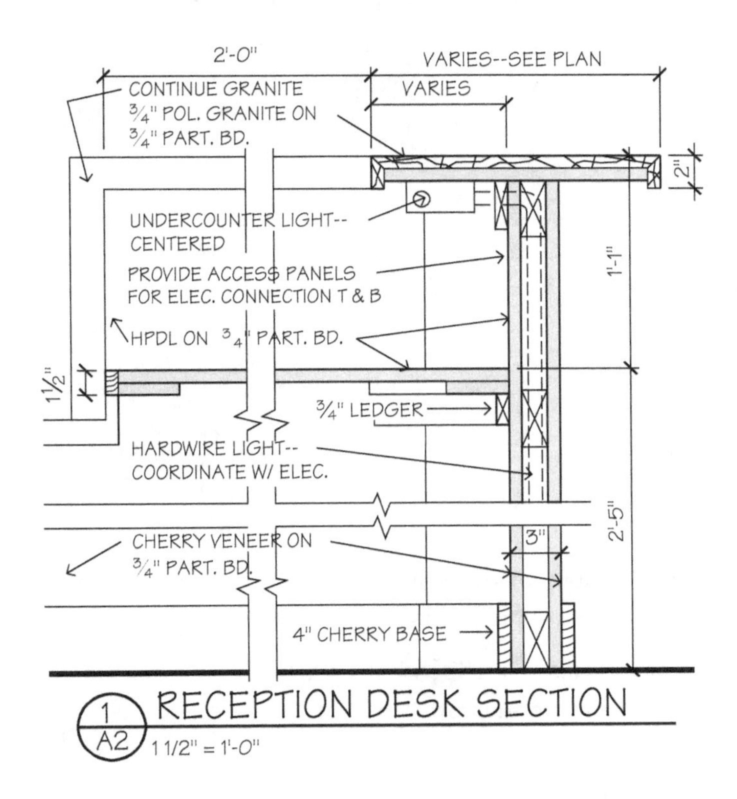

31. After creating the reception desk section detail, the interior designer must consult with the

(A) electrical engineer
(B) building owner
(C) furniture dealer
(D) mechanical engineer

32. Who would be responsible for ordering and placing a stone top of a custom-built reception desk that includes woodwork, stone, metal, and a light fixture?

(A) architect
(B) interior designer
(C) stone fabricator
(D) woodwork fabricator

33. Who is responsible for verifying that recessed downlights do not interfere with ductwork shown on the mechanical plans?

(A) mechanical engineer
(B) electrical engineer
(C) interior designer
(D) architect

34. The final responsibility of awarding a contract rests with the

(A) interior designer
(B) construction manager
(C) owner
(D) owner's legal counsel

35. An interior designer has written furniture specifications for a project and has put the job out for bid to several dealerships. In this situation, purchase orders are MOST likely to be written by the

(A) owner
(B) dealership
(C) furniture rep
(D) interior designer

36. Who is responsible for determining the location and general condition of existing light fixtures as part of a due diligence investigation?

(A) electrical engineer
(B) electrical subcontractor
(C) general contractor
(D) interior designer

37. Which type of chart is MOST often used with complex projects to show critical tasks and their associated times?

(A) bar chart
(B) CPM chart
(C) Gantt chart
(D) PERT chart

38. One of the primary advantages of a full wall schedule is that it

(A) takes less time to develop than a CPM chart
(B) is more detailed than a Gantt chart
(C) is large enough for several people to view at once
(D) involves the entire project team in the scheduling

39. The interior designer is responsible for which of the following aspects of consultant coordination? (Choose the four that apply.)

(A) giving the electrical engineer information about the client's equipment
(B) providing the mechanical engineer with the occupant load for each space
(C) informing the mechanical engineer about current sprinkler system codes
(D) making sure that the final drawings and specifications are coordinated
(E) telling the electrical engineer where to place the electrical closets
(F) giving the electrical engineer the designer's preferred choices for luminaires

40. When coordinating with a security consultant, the interior designer's drawings should show

(A) a schedule of all security devices
(B) the emergency backup power supply
(C) the positioning of required lighting
(D) the wiring of the security devices

41. Thermostat locations are determined by the

(A) architect
(B) electrical engineer
(C) HVAC contractor
(D) mechanical engineer

42. As construction work begins on a large commercial project, the interior designer is concerned about coordinating the installation of services in the plenum so that sufficient space is allowed for planned recessed lighting and raised ceilings. Which type of subcontractor should the designer have the general contractor coordinate with first?

(A) electrical
(B) mechanical
(C) plumbing
(D) sprinkler

43. An interior designer can determine what products an independent representative is handling by consulting

(A) the manufacturer's website
(B) a showroom directory
(C) Sweets Network Online
(D) a line chart

44. An interior designer wants to specify acoustic insulation. Which type of specification should be used?

(A) performance
(B) prescriptive
(C) proprietary
(D) reference

45. An interior designer recommends that a client purchase appliances directly rather than through a general contractor. What would be the primary reason for this recommendation?

(A) The client could avoid the contractor's markup.
(B) The client could not be overcharged for delivery and installation.
(C) The interior designer could provide the client a broader selection by acting as the client's agent.
(D) The client could get a better discount than the contractor.

46. Under the provisions of AIA Document B152, *Standard Form of Agreement Between Owner and Architect for Interior Design and Furniture, Furnishings, and Equipment (FF&E) Design Services*, which of the following are part of the owner's responsibilities? (Choose the four that apply.)

(A) arranging for laboratory tests required by the contract documents
(B) removing existing furniture from the project site
(C) providing space for the receipt and storage of materials used on the project
(D) giving copies of the drawings to the contractor
(E) providing a written program for the project
(F) preparing the bidding documents

47. Under the provisions of AIA Document B152, *Standard Form of Agreement Between Owner and Architect for Interior Design and Furniture, Furnishings, and Equipment (FF&E) Design Services*, which of the following are among the interior designer's basic services? (Choose the three that apply.)

(A) inspecting and accepting or rejecting furniture
(B) reviewing laws and regulations applicable to the project
(C) developing a schedule of milestone dates when the owner's decisions must be made
(D) maintaining a record of changes relative to the work
(E) conducting an existing FF&E inventory
(F) appearing at legal proceedings related to the project

48. Which of the following are advantages of negotiated contracts? (Choose the four that apply.)

(A) The client can select the contractor based on criteria other than price.
(B) The contractor will provide the highest-quality project.
(C) The contract price will be lower than it would be under other contract types.
(D) The schedule is generally shorter than it would be under other contract types.
(E) The interior designer is involved in the process.
(F) The client can select the subcontractors.

49. Which of the following are typically part of an interior designer's post-occupancy evaluation? (Choose the four that apply.)

(A) evaluating the HVAC system
(B) interviewing representative users of the project
(C) suggesting how the client's next project can be improved
(D) reviewing construction costs with the client
(E) checking maintenance problems
(F) reviewing the performance of the contractors

50. Tracking, or monitoring, a project during design and construction, whether by hand-made chart or computer program, involves

(A) using time sheets to measure productivity and minimum billing rates
(B) comparing estimated fees and time against actual expenditures and percent complete
(C) comparing client expectations at regular intervals during construction against original goals
(D) verifying actual construction costs against the estimated budget as the project is completed

51. A developer is planning to build a 50-unit multi-family housing project with an interior designer providing services for the design and construction documents for interior spaces. The developer wishes to encourage competitive bidding but also wants to ensure that the materials used meet a minimum standard of quality. Which type of specification should be used?

(A) proprietary
(B) base bid with "approved equal" language
(C) descriptive
(D) base bid with alternates

52. What are the basic goals of a quality control program during the design phase of a project? (Choose the three that apply.)

(A) using the best employees for the project
(B) continually meeting the client's original goals and objectives
(C) designing an award-winning project
(D) providing the standard of care for professional practice
(E) meeting the expectations of the future users of the project
(F) reducing risks and liability

❸ PROFESSIONAL BUSINESS PRACTICES

53. Which of the following is true of title acts versus practice acts regulating the practice of interior design?

(A) practice acts require meeting educational requirements and passing the NCIDQ exam
(B) states with title acts require continuing education to keep their license current
(C) the allowable titles under title laws are uniform for each state
(D) practice acts do not allow the use of the title "interior decorator"

54. The client is MOST typically involved with procurement of furniture through

(A) the acknowledgment
(B) the freight bill
(C) a purchase order
(D) a sales agreement

55. An interior designer could determine which clients have not paid by looking at the

(A) aged accounts receivable
(B) balance sheet
(C) cash flow statement
(D) income statement

56. Which type of accounting would MOST likely be chosen by an interior designer working alone as a sole proprietorship?

(A) accrual
(B) cash
(C) double-entry
(D) ledger-based

57. Which of the following provides the MOST accurate basis for estimating project costs?

(A) square footage
(B) functional unit
(C) parameter
(D) quantity takeoff

58. Which of the following is the interior designer's BEST source for determining a piece of furniture's net price?

(A) showroom manager
(B) manufacturer's price list
(C) furniture dealership
(D) representative's line chart

59. According to AIA Document B152, *Standard Form of Agreement Between Owner and Architect for Interior Design and Furniture, Furnishings, and Equipment (FF&E) Design Services,* the interior designer must prepare cost estimates for the project during the

(A) schematic design phase only
(B) design development phase only
(C) schematic design and design development phases
(D) schematic design, design development, and construction documents phases

60. To avoid liens against a project, the interior designer should suggest that the owner require the contractor to submit a

(A) bid bond
(B) bid security
(C) performance bond
(D) labor and material payment bond

61. The designer for the interior furnishings of a large public hospital project financed with bond money has assisted the city government in preparing the bidding documents. When the bids from five qualified contractors are opened, they are all over budget, ranging from 4% to 10% over the approved costs. What should the designer do?

(A) Recommend that the city accept the lowest bid and obtain the extra 4% from other sources.
(B) Begin to study ways to reduce the projects scope so it meets the budget.
(C) Suggest that the project be rebid because the lowest bid is so close to the budget.
(D) Wait for the city to tell the designer how it wants to proceed.

62. The requirements for how bidders should propose a substitution is found in the

(A) advertisement to bid
(B) bidding procedures
(C) instructions to bidders
(D) general conditions

63. Which of the following is used to accommodate costs that are not yet confirmed at the time of a bid?

(A) allowance
(B) add-alternate
(C) material bond
(D) unit price

64. Bids have been submitted by four contractors on a midsize restaurant. The lowest bid is 10% over the client's budget, and the next lowest bid is 12% over budget. What is the interior designer's BEST course of action?

(A) Suggest that the client obtain additional financing for the extra 10%, and accept the lowest bid.
(B) Accept the lowest bid, and tell the contractor that in order to get the job, the contractor must reduce the bid by 10%.
(C) Work with the client to redesign the project to reduce the cost.
(D) Remind the client that the designer is not responsible for construction costs, and tell the client that additional money is needed.

65. Which of the following are typically considered part of the FF&E contract? (Choose the four that apply.)

(A) a commissioned sculpture bolted to a wall
(B) vending machines built into an opening
(C) auditorium seating
(D) wall-to-wall carpeting
(E) applied acoustical panels
(F) vertical blinds

66. A project construction budget prepared by the interior designer would include which of the following line items? (Choose the four that apply.)

(A) designer's fees and anticipated reimbursable expenses
(B) estimates for built-in equipment
(C) fixtures to be supplied by the owner
(D) contractor's profit
(E) custom chandeliers installed by the contractor
(F) allowances for custom area rugs

67. An interior designer's drawings indicate an area of hardwood flooring with an indeterminate boundary because the client is still deciding on the flooring's actual area. To ensure that bid prices for the flooring are compared fairly, which of the following should the interior designer request be included on the bid form?

(A) individual quotes
(B) unit prices
(C) fixed costs
(D) contingencies

68. The risk for furniture being damaged during shipment is assigned by the

(A) FF&E General Conditions
(B) Owner-Designer Agreement
(C) Owner-Vendor Agreement
(D) *Uniform Commercial Code*

69. Under AIA Document A151, *Standard Form of Agreement Between Owner and Vendor for Furniture, Furnishings, and Equipment (FF&E),* the inspection and acceptance of furniture on delivery to the jobsite is the responsibility of the

(A) dealership
(B) furniture rep
(C) interior designer
(D) owner

70. A transaction privilege tax license allows an interior design firm to

(A) conduct business
(B) buy and sell furniture
(C) bill clients for the cost of employees
(D) charge clients for installation services

4 CODE REQUIREMENTS, LAWS, STANDARDS, AND REGULATIONS

71. In addition to meeting the requirements of the ADA, what other accessibility requirements should the interior designer be MOST concerned with when doing design development for a commercial project?

(A) ANSI A117.1, *Accessible and Usable Buildings and Facilities*
(B) scoping provisions of the local building code
(C) *ADA/ABA Guidelines*
(D) *Uniform Federal Accessibility Standards*

72. Most building codes in the United States are established by

(A) federal laws
(B) model code-writing agencies
(C) state governments
(D) local governments

73. ASTM is a(n)

(A) model code group
(B) industry standard-writing organization
(C) testing laboratory
(D) federal code-writing agency

74. When working under the requirements of the *International Energy Conservation Code* (IECC), the interior designer will MOST likely use which method to assist in the lighting design of a commercial project?

(A) building area method
(B) energy cost budget method
(C) luminaire efficacy method
(D) space-by-space method

75. In order to specify an acceptable type of wall covering, which of the following tests should the designer require that the wall covering pass?

(A) methenamine pill test
(B) smoke density chamber test
(C) Steiner tunnel test
(D) vertical ignition test

76. A building that carries a gold rating has been designed and certified under which of the following systems?

(A) Greenguard
(B) Green Seal
(C) ISO 14000
(D) LEED

77. A client has approached an interior designer to remodel a large mansion into a restaurant and nightclub. Before accepting the commission or beginning any work, with whom should the designer consult?

(A) building department
(B) city planning commission
(C) department of excise
(D) zoning department

78. If a material does not burn, it is considered to be

(A) fire retardant
(B) fire rated
(C) flame resistant
(D) noncombustible

79. Which of the following organizations or standards establishes volatile organic compound (VOC) limits on indoor air quality for furniture systems?

(A) BIFMA
(B) EPA
(C) IBC
(D) IECC

80. An interior designer is evaluating various alternatives for a particular building product as they relate to sustainability. The interior designer would MOST likely use a(n)

(A) environmental impact study
(B) life-cycle assessment
(C) impact assessment
(D) matrix comparison chart

81. To verify a manufacturer's claim that its product has met the requirements of an ASTM standard, the interior designer should

(A) contact the relevant trade association for the product
(B) contact ASTM and get a copy of the standard
(C) investigate the standard on internet sites
(D) request detailed information from the manufacturer

82. What testing standard should be specified for the durability of office chair construction?

(A) ANSI/BIFMA X5.1
(B) ASTM D4157
(C) CAL TB133
(D) NFPA 701

83. Which of the following energy conservation standards is the interior designer most likely to use when designing a commercial project?

(A) IgCC
(B) IBC
(C) IECC
(D) IRC

84. When is the application for a commercial building permit submitted?

(A) when the construction drawings are 95% complete
(B) when the interior designer has completed the construction documents
(C) after the contractor has approved the final construction documents
(D) when the owner has received the final copy of the drawings and specifications

85. A client has asked an interior designer to work on a new project that will involve changing the use of an existing building from a four-plex residence to an artists' cooperative workspace. What agency must the designer first consult to determine if this would be possible?

(A) building department
(B) city council
(C) planning commission
(D) zoning department

86. An interior designer beginning a multifamily housing project would want to consult the

(A) *ADA Standards for Accessible Design*
(B) Fair Housing Act
(C) *International Building Code*
(D) *Uniform Federal Accessibility Standards*

87. Where does a common path of egress travel end?

(A) in an area of refuge
(B) at the exterior exit from the building
(C) at the point where an individual has a choice about which direction to go to reach an exit
(D) at a public way

88. According to model codes, which of the following are considered parts of the means of egress? (Choose the three that apply)

(A) common path of travel
(B) exit
(C) exit access
(D) exit discharge
(E) public way
(F) travel distance

89. An electrical engineer has informed the interior designer that the lighting in a series of meeting rooms in their project will, by code, be limited to 0.82 W/ft^2 and the designer will have to select luminaires accordingly. This is known as

(A) an energy cost budget
(B) a space-by-space method
(C) a lighting power density
(D) a lighting power allowance

90. In addition to the ability to pass the hose stream test and size limitations, what other factors must be considered when selecting fire-resistive-rated glazing for a corridor in an interior project?

(A) impact rating and requirements for a heat barrier
(B) certification of the manufacturer and impact rating
(C) the direction of the expected fire attack
(D) requirements for a heat barrier and certification of the manufacturer

91. In order for glazing to be considered fire-resistive-rated, which of the following reference standards must it comply with?

(A) 16 CFR 1201
(B) ASTM E84
(C) ASTM E119
(D) NFPA 701

92. In order to get a permit from the authority having jurisdiction (AHJ), what type of plan must the interior designer develop for an office project that occupies a portion of a full floor in a high-rise building?

(A) furniture plan
(B) location plan
(C) power plan
(D) sprinkler plan

93. In order to easily understand what types of visible and audio alarms are required for a project and where they are required, what reference would the interior designer find most useful and easy to use?

(A) *International Building Code*
(B) *International Fire Code*
(C) *National Electrical Code*
(D) NFPA *National Fire Alarm and Signaling Code*

94. While there are minor variations in accessibility standards, which reference standard are they all based on?

(A) 2010 *ADA Standards for Accessible Design*
(B) 28 CFR 36
(C) *ADA/ABA Guidelines*
(D) ICC A117.1

95. According to the *International Energy Conservation Code*, which of the following are required and would be most appropriate in an office suite with individual offices?

(A) daylight-responsive controls
(B) occupant sensor controls
(C) time-switch controls
(D) toplight daylight zone controls

96. Which of the following is most likely to be found in a local zoning ordinance?

(A) minimum setbacks from property lines
(B) maximum occupancy
(C) fire-rated assemblies
(D) types of material that may be used for construction

97. During the design of a restaurant, the interior designer finds that a local building code amendment conflicts with state health department requirements. What action should the interior designer take?

(A) Appeal to the health department.
(B) Ask for a ruling from the building department.
(C) Have the client resolve the issue.
(D) Design to the most restrictive requirement.

98. Which of the following are important criteria for evaluating the selection of building materials with the goal of conserving resources? (Choose the four that apply.)

(A) recycled content
(B) maintainability
(C) potential for reuse
(D) embodied energy
(E) low toxicity
(F) amount of renewable raw materials

99. What reference standard should be used to determine the most complete and detailed requirements for sprinkler spacing and positioning?

(A) IBC
(B) IPC
(C) IMC
(D) NFPA

100. The design occupant load of a hotel ballroom is determined by

(A) drawing proposed layouts of the room for reception, meeting, and dining, and figuring out how many chairs will fit into the space in each scenario
(B) dividing the total area of the space by the area required per person
(C) multiplying the number of people expected to use the space by an assumed weight per person
(D) asking the conference services manager how many people must be accommodated in the space

101. What organization establishes the most referenced standards that interior designers use for commercial office furniture?

(A) ANSI
(B) ASTM
(C) BIFMA
(D) NFPA

102. A restaurant has a maximum occupancy of 300 people. For all exits, the building code requires an allowance of 0.2 in per occupant. Calculate the minimum number and size of exits.

(A) one exit, 5 ft 0 in in pair of doors
(B) one exit, 6 ft 0 in in pair of doors
(C) two exits, two 30 in doors
(D) two exits, two 36 in doors

103. What is the minimum size for a shower compartment, not including exceptions?

(A) 24 in by 30 in (610 mm by 762 mm)
(B) 30 in by 30 in (762 mm by 762 mm)
(C) 30 in by 36 in (762 mm by 914 mm)
(D) 36 in by 36 in (914 mm by 914 mm)

104. The number of required exits from a space depends on which of the following? (Choose the four that apply.)

(A) common path of egress travel
(B) maximum travel distance
(C) the occupancy of the space
(D) exit or exit access width
(E) the occupant load
(F) if the building is sprinklered

105. A local code official has the authority to enforce all but which of the following codes or acts?

(A) Americans with Disabilities Act (ADA)
(B) *Life Safety Code* (NFPA 101)
(C) *International Building Code* (IBC)
(D) ICC/ANSI A117.1

❺ INTEGRATION WITH BUILDING SYSTEMS AND CONSTRUCTION

106. Under what conditions would a $1^3/_4$ in (44.4 mm) thick hollow metal door be used?

(A) when a fire rating of over 90 minutes and high security and durability are required
(B) when a fire rating over 1 hour is required or a steel frame is necessary
(C) when the door opening is expected to receive heavy use and a smoke-proof opening is required
(D) when the door will receive minimal maintenance under heavy use

107. Safety glazing is required in which of the following? (Choose the three that apply.)

(A) glass panels where the sill is greater than 18 in (457 mm) above the floor
(B) glass sidelights next to a solid wood door
(C) shower doors
(D) full-height glass panels less than 12 in (305 mm) from a door
(E) full-height glass panels with a conforming crash bar
(F) glass openings less than 9 ft^2 (0.84 m^2) in area

108. When selecting and detailing window coverings for a building, what is the MOST important consideration to coordinate with the mechanical engineer?

(A) degree of transparency of the covering
(B) direction of airflow from the ceiling toward the window
(C) distance between the window and the window covering
(D) color of the window covering

109. What is the interior designer's BEST course of action if a client wants to remove part of a partition thought to be a bearing wall?

(A) have a structural engineer or architect review the problem and make a recommendation
(B) look in the attic to see if the wall supports any beams or joists
(C) ask the client for the building's structural drawings and review them
(D) check for structural stability by tapping on the wall and determining where it is located in relation to the center of the building

110. A client has requested that special security protection be provided at a reasonable cost for critical paper files containing corporate trade secrets. In designing the file room, the interior designer should suggest

(A) card readers at the doors leading to the file room
(B) photoelectric beams within the room
(C) magnetic contacts on all doors leading to the file room
(D) electronic shielding of the file room

111. If all of the following are present, what MUST be modified to achieve acoustic separation in a perimeter office?

(A) acoustical tiles
(B) gypsum wallboard
(C) convector
(D) batt insulation

112. During the design development phase of planning a large classroom, which of the following methods could be used to reduce potential noise problems in a room of a given area whose exact shape has not yet been determined? (Choose the four that apply.)

(A) Design the room to have the largest ceiling surface area possible.
(B) Minimize the length of the wall separating the classroom from the noisiest adjacent room.
(C) Plan to install sound-absorbent material on the walls of the classroom.
(D) Study ways to increase the transmission loss of the room's partitions.
(E) Specify a white noise system for the room.
(F) Locate the doors to the classroom away from doors of nearby rooms.

113. A device used to protect against fires from faulty wiring is a(n)

(A) arc-fault interrupter
(B) circuit breaker
(C) disconnect switch
(D) ground fault interrupter

114. What is the SAFEST type of glazing to use in a sidelight adjacent to a door?

(A) ceramic
(B) wire
(C) float
(D) tempered

115. Core drilling would be MOST difficult in which type of structural system?

(A) concrete over steel deck
(B) flat slab concrete
(C) one-way pan joist concrete
(D) post-tensioned concrete

116. The interior designer can design details to accommodate building movement caused by which of the following forces? (Choose the three that apply.)

(A) building expansion
(B) floor deflection
(C) foundation settlement
(D) seismic events
(E) thermal expansion
(F) wind sway

117. A residential client with a one-story house wants to remove an 8 ft (2438 mm) section of a loadbearing wall to make an opening between two existing rooms. What is the MOST appropriate advice the interior designer can give?

(A) The opening can be framed, but a structural engineer will be needed.
(B) An architect will have to be retained to sign the drawings for the building department.
(C) The opening can be framed with a double 2 in × 8 in (38 mm × 184 mm) header.
(D) The proposed opening cannot be made because the wall is a load-bearing wall.

118. Which type of structural system in an existing building would allow for the easiest penetration for a new stair between two floors?

(A) beam and girder steel
(B) flat plate concrete
(C) hollow-core slab
(D) open web steel joists

119. What type of system would be BEST for an open office plan so the heating and cooling for each workstation could be individually controlled?

(A) all-air system
(B) all-water system
(C) radiant panel system
(D) air-water system

120. What type of sprinkler head should be used for a decorative, open-grid, wood-slat ceiling suspended from the structural floor above?

(A) upright
(B) pendent
(C) sidewall
(D) recessed

121. The location of a private toilet room on the upper floors of a high-rise building is MOST limited by

(A) length of the hot and cold supply lines
(B) floor-to-floor dimensions
(C) distance from an existing drain
(D) plenum dimension

122. On a large commercial project, the location of telephone and computer outlets is determined by the

(A) building architect
(B) electrical engineer
(C) interior designer
(D) telecommunications consultant

123. The location of a new toilet room in an existing building would be MOST limited by the

(A) location of cold water pipes
(B) position of existing vent stacks
(C) location of an overhead structural beam
(D) distance from the nearest soil stack

124. Which of the following strategies provides the MOST effective above-ceiling acoustical separation between two adjacent rooms, assuming all cracks between materials are suitably sealed?

(A) ceiling tiles with a high ceiling attenuation class (CAC) rating
(B) sheet lead draped from the slab above to the top of the ceiling tile
(C) partitions extended through the ceiling to the slab above
(D) framing suspended from the slab to the ceiling, with one layer of gypsum board attached on one side

125. Which type of structural system would allow for the highest finished ceiling?

(A) flat plate
(B) flat slab
(C) open web steel joist
(D) waffle slab

126. When a double door with an astragal is included in a fire separation partition, which of the following is required?

(A) automatic door bottom
(B) door coordinator
(C) panic hardware
(D) threshold

127. Air is most effectively distributed along an exterior window wall by a

(A) round diffuser
(B) square grille
(C) series of square diffusers
(D) slot air diffuser

128. The partition type shown in the illustration is typically used when

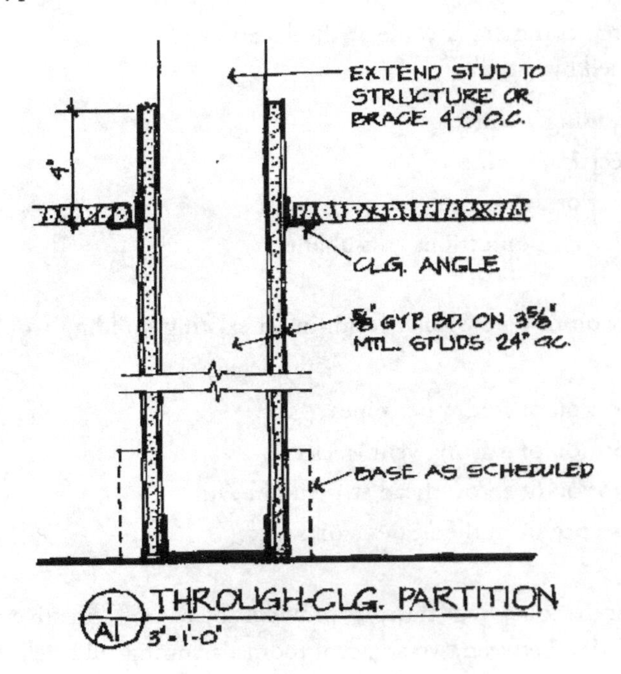

(A) additional sound control is needed
(B) shelving will be installed on one side
(C) suspended ceilings are at different heights
(D) earthquake bracing is required

129. In creating drawings for paneling that will be suspended from a wall with cleats, the MOST important piece of information is the

(A) width of each panel along the wall
(B) dimension between the panel top and ceiling
(C) thickness of the wood cleat
(D) size of the base

130. The detail shown in the illustration is an example of

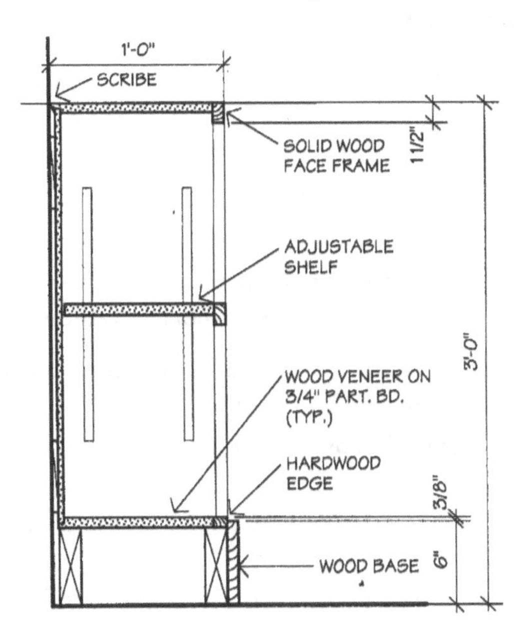

(A) architectural woodwork
(B) finish carpentry
(C) manufactured casework
(D) standard cabinetry

131. Commercial-grade cabinets are MOST often constructed of panel products with a thickness of

(A) $^1/_2$ in (13 mm)
(B) $^5/_8$ in (16 mm)
(C) $^3/_4$ in (19 mm)
(D) $^7/_8$ in (22 mm)

6 INTEGRATION OF FURNITURE, FIXTURES, AND EQUIPMENT

132. Furniture budgets can be accurately estimated by the

(A) general contractor and furniture manufacturer
(B) interior designer and furniture dealers
(C) furniture representative and client
(D) furniture manufacturer and interior designer

133. When the interior designer does not want furniture to be delivered to the designer's office, the designer should complete a

(A) customized purchase order
(B) drop ship order
(C) letter of intent
(D) tag for label

134. What would be the BEST way to ensure that the finish on new millwork matches the finish on existing millwork on a remodeling project?

(A) Indicate on the drawings and in the specifications that the new work should match the existing work.
(B) Ask the client to find out what was used on the old job and include that information in the specifications.
(C) Research the manufacturer and color of the existing finish and include that information in the specifications.
(D) Ask the painting contractor to investigate what finish was previously used and include that information in the finish schedule.

135. Which of the following sources would be MOST appropriate for the warehouse portion of a large furniture dealership?

(A) high-pressure sodium
(B) cool-white deluxe fluorescent
(C) metal-halide
(D) mercury-vapor

136. In preparing an FF&E budget, where would the interior designer get the MOST accurate cost information on a sofa?

(A) furniture manufacturer
(B) furniture dealer
(C) internet
(D) merchandise mart

137. A designer is beginning to lay out an office space that will be based on open planning and flexibility. What type of furniture would be most appropriate for this use?

(A) ready-made
(B) moveable panels with ancillary pieces
(C) systems furniture
(D) custom-designed

138. When designing and specifying acoustic panes for sound control, what are the most important considerations? (Choose the four that apply.)

(A) permeable fabric
(B) hydrophobic fabric
(C) loose core material
(D) tackable core material
(E) fabric color
(F) unbacked fabric

139. Demountable partitions systems offer which of the following advantages. (Choose the four that apply.)

(A) smooth finish along panels
(B) flexible layout
(C) reuse potential
(D) low life-cycle costs
(E) complete system with door frames and similar components
(F) low initial cost

140. When developing a furniture schedule, which of the following should be included for a complete schedule? (Choose the four that apply.)

(A) keynote number
(B) description
(C) size
(D) manufacturer's catalog number
(E) color
(F) manufacturer's name

141. An interior designer wants to select woven upholstery for a client's sofa that will not quickly need repair or replacement. The interior designer should specify that the upholstery to be selected must be classified as "heavy duty" on the

(A) Martindale abrasion test
(B) Taber abraser test
(C) tearing strength test
(D) Wyzenbeek abrasion resistance test

142. What is the most important human factor to consider when designing display fixtures for a jewelry store?

(A) anthropometrics
(B) ergonomics
(C) gestalt
(D) proxemics

143. An interior designer is developing plans and details for custom layout tables in a quilt shop. The tables will be used by the clerks to help customers look at patterns, view fabrics, and compare fabric colors. When designing these layout tables, which of the following accessibility issues must the interior designer take into consideration? (Choose the four that apply.)

(A) clear space around the tables
(B) table height
(C) knee space
(D) equivalent facilitation
(E) reach ranges
(F) tolerances

144. In the furniture procurement process, the interior designer gives the exact requirements of what the client has selected with the

(A) sales agreement
(B) acknowledgment
(C) purchase order
(D) freight billing

145. Of the various ways an interior designer can procure furniture for a client, which is the riskiest and requires a resale license?

(A) acting as a purchasing agent for the client
(B) by writing furniture specifications and giving them to a dealership
(C) ordering the furniture and hiring an expediter to complete delivery and installation
(D) completing all aspects of ordering, installing, and paying for the furniture

146. On projects that involve both construction and furniture, fixtures, and equipment (FF&E), why are FF&E typically separate budget items from construction?

(A) they are specified, purchased, and installed differently than construction
(B) they are purchased wholesale by the interior designer with a markup applied
(C) most clients maintain separate monetary accounts for FF&E purchases
(D) they contain things like artwork, rugs, and window coverings that the contractor can't estimate

147. What is attic stock?

(A) excess items that are leftover when too much has been ordered for a project
(B) extra materials or furniture components purchased under contract and stored for later use
(C) perishable materials, like paint, that require special storage conditions
(D) high-cost furniture items that may be needed for future facility expansion

148. In addition to the original cost of a large number of furniture items, what costs should the interior designer consider when performing a life-cycle cost analysis (LCC)? (Choose the four that apply.)

(A) maintenance costs over the length of the study period at present value
(B) operational costs discounted to the base date
(C) the future cost of any replacements at the time of replacement
(D) taxes over the length of the study period
(E) finance costs at the end of the study period
(F) any residual value at the end of the study period

149. Which of the following are typically part of the furniture, fixtures, and equipment (FF&E) budget? (Choose the four that apply.)

(A) appliances
(B) lamps
(C) chandeliers
(D) wood cabinets
(E) rugs and mats
(F) window coverings

❼ CONTRACT ADMINISTRATION

150. When the project is complete and the final inspection has been completed, what allows the client to take possession of the space?

(A) certificate of occupancy
(B) final certificate for payment
(C) final commissioning report
(D) punch list

151. Which document provides the designer with verification that an item was ordered correctly?

(A) acknowledgment
(B) bill of lading
(C) packing slip
(D) purchase order

152. A project is being completed subject to the standard AIA A201, *General Conditions of the Contract for Construction* agreement. If some millwork was installed with the incorrect finish and no sample was submitted to the designer, who is responsible?

(A) millworker
(B) interior designer
(C) project manager
(D) contractor

153. If an interior designer missed an error in a dimension on the shop drawings for a custom steel doorframe and the doorframe was subsequently fabricated, who is responsible for paying to have the mistake corrected?

(A) door frame supplier
(B) wall framing subcontractor
(C) general contractor
(D) interior designer

154. A doorway is installed by the contractor according to the drawings. After viewing the job, a building inspector tells the contractor that the door is not wide enough. Who is responsible for correcting the problem?

(A) framing subcontractor
(B) interior designer
(C) owner
(D) contractor

155. If, during construction of a project, the contractor notices that a handrail does not meet the local building codes, what action could the designer reasonably expect the contractor to take?

(A) The contractor should notify the designer of the discrepancy in writing.
(B) The contractor should correct the situation and submit a change order for the extra work.
(C) The contractor should build the handrail according to the contract documents because conformance to building codes is the designer's responsibility.
(D) The contractor should notify the designer of the problem and suggest a remedy.

156. Which of the following are TRUE about addenda? (Choose the four that apply.)

(A) An addendum is sent only to the contractor requesting the clarification.
(B) An addendum is issued only before the construction contract is signed.
(C) An addendum is used to modify or interpret the contract documents.
(D) An addendum must be filed with other bidding documents on file in a plan room.
(E) An addendum must be issued no later than four days before bid opening.
(F) An addendum can be issued after the construction contract is signed.

157. Bid openings are typically attended by which of the following parties? (Choose the three that apply.)

(A) owner
(B) interior designer
(C) material suppliers
(D) contractor
(E) subcontractors
(F) lending agent

158. Two weeks before the bids are due on a large restaurant project, one of the contractors asks if he can price a type of ceiling tile that was not listed in the specifications. What action should the interior designer take?

(A) Advise the contractor that he should submit backup proof with the bid that the proposed change is equal to what was specified.

(B) Refer the contractor to the owner, who will make the final determination, and then notify the other bidders that this has been done.

(C) Tell the contractor to request approval in writing.

(D) Issue an addendum stating that one of the contractors has asked for permission to price an alternate and that all contractors may do this.

159. Which of the following are among the responsibilities of the project manager? (Choose the four that apply.)

(A) hiring new staff to work on a project

(B) reviewing office-wide aged accounts receivables

(C) verifying that office contractual obligations are being met

(D) reviewing billings from the project's consultants

(E) assisting in writing proposals for a specific project

(F) coordinating activities of consultants

160. During a project's planning stage, a project manager is responsible for

(A) monitoring and directing

(B) scheduling and coordinating

(C) coordinating and monitoring

(D) fee projecting and time planning

161. The project manager should document

(A) only those items that belong in the project notebook

(B) only standard forms and written correspondence

(C) only those documents that have legal significance

(D) all items relevant to a project and its administration

162. The project manager is responsible for which of the following schedules? (Choose the four that apply.)

(A) design schedule

(B) bidding schedule

(C) construction schedule

(D) furniture schedule

(E) predesign programming schedule

(F) commissioning schedule

163. A Gantt chart is another name for a

(A) bar chart
(B) full wall chart
(C) critical path chart
(D) PERT chart

164. Which of the following have the MOST influence on how quickly a project is completed in the designer's office? (Choose the four that apply.)

(A) client's approval process
(B) client's staff size
(C) design firm's staff size
(D) agency approval
(E) project complexity
(F) design philosophy of design offices' senior staff

165. Which of the following are correct concerning following up with a project after it has been completed? (Choose the three that apply.)

(A) It is required in standard contracts as a basic service.
(B) It provides valuable information for future projects.
(C) Professional photography of the project is required for the client.
(D) It makes a favorable impression on a client.
(E) It helps solve minor problems after move-in.
(F) Post-occupancy evaluation should be undertaken within three months.

166. Which of the following are MOST likely to be part of the project manager's responsibilities? (Choose the four that apply.)

(A) keeping notes on daily decisions and meetings
(B) organizing the layout of construction drawings
(C) generating strategic, long-term goals for the firm
(D) planning and assigning weekly job tasks for the project staff
(E) monitoring fee expenditures throughout all phases of the project
(F) staying current with the client's opinion of the progress of the project

167. If an interior designer specifies file cabinets that do not fit within a space the contractor built according to the contract documents, who is responsible for paying for the correction?

(A) contractor
(B) interior designer
(C) cabinet supplier
(D) owner

168. What document is used to approve the release of funds for furnishing a project?

(A) application for payment
(B) certificate for payment
(C) purchase order
(D) bill of lading

169. The owner is protected from incomplete work by the contractor by the use of

(A) indemnification
(B) liquidated damages
(C) retainage
(D) standard contract forms

170. Which of the following are considered vital for project meetings? (Choose the four that apply.)

(A) appropriate participants based on the phase of the project
(B) notes with a complete and unbiased account of the topics of discussion
(C) a statement on the notes that they represent the final results of the meeting
(D) notes taken only in a handwritten, contemporaneous form for later printing
(E) distribution of recordings of the project meetings to the client
(F) timely distribution of meeting notes to those attending

171. When a mock-up is required by the interior designer's specifications, who is responsible for paying for the mock-up if it does not meet the requirements?

(A) general contractor
(B) interior designer
(C) millwork subcontractor
(D) owner

172. A millwork subcontractor drops off a box of countertop material and cabinet hardware samples at the interior designer's office with a transmittal requesting selections by the end of the week. What should the interior designer do with the materials?

(A) Review the options and call the subcontractor with the selection.
(B) Review the options, make the selections, and inform the contractor in writing of the choices.
(C) Send the samples back to the subcontractor.
(D) Send the samples to the contractor.

173. During the construction of a hotel lobby, the client decides to replace the porcelain tile floor specified for the floor with a poured terrazzo flooring material. The interior designer requests a proposal from the contractor for the change. The contract states that the contractor may add 20% for overhead and profit and an additional 5% for coordination on change orders. The contractor's base price for labor and materials for the change is $22,225. What is the approximate total value for the change order?

(A) $26,000
(B) $27,000
(C) $28,000
(D) $29,000

174. Which of the following is used to formally incorporate a substitution into the work prior to award of the contract?

(A) change order
(B) addendum
(C) alternate listing
(D) construction change directive

175. Which of the following requires on-site approval?

(A) mock-ups
(B) product data
(C) samples
(D) shop drawings

MOCK EXAM ANSWERS

Once you have completed your mock exam, compare your answers to those on the answer key that follows. The explanations to these answers begin on the page following the answer key.

❶ Project Assessment and Sustainability

1. B
2. A
3. C
4. A
5. B, C, E, F
6. B
7. D
8. B
9. B
10. B, D, E, F
11. *See Sol. 11*
12. B, C, E, F
13. A, B, D, E
14. D
15. C
16. B, C, D
17. A
18. 10,000 ft^2 (900 m^2)
19. A, C, D, F
20. D
21. B
22. B
23. C
24. A
25. A, B, C, D
26. D

❷ Project Process, Roles, and Coordination

27. C
28. A, C, E, F
29. D
30. B
31. A
32. D
33. C
34. C
35. B
36. D
37. B
38. D
39. A, B, D, F
40. C
41. D
42. B
43. D
44. D
45. A
46. A, B, C, E
47. B, C, D
48. A, D, E, F
49. A, B, E, F
50. B
51. B
52. B, D, F

❸ Professional Business Practices

53. A
54. D
55. A
56. B
57. D
58. C
59. D
60. D
61. B
62. C
63. A
64. C
65. A, B, C, F
66. B, C, D, E
67. B
68. D
69. D
70. B

❹ Code Requirements, Laws, Standards, and Regulations

71. B
72. D
73. B
74. D
75. C
76. D
77. D
78. D
79. A
80. B
81. D
82. A
83. C
84. B
85. D
86. B
87. C
88. B, C, D
89. C
90. A
91. C
92. B
93. A
94. D
95. B
96. A
97. D
98. A, C, D, F
99. D
100. B
101. C
102. D
103. B
104. A, C, E, F
105. A

❺ Integration with Building Systems and Construction

106. A
107. B, C, D
108. C
109. A
110. A
111. C
112. B, C, D, F
113. A
114. D
115. D
116. B, D, F
117. A
118. A
119. A
120. A
121. C
122. C
123. D
124. C
125. A
126. B
127. D
128. C
129. B
130. A
131. C

❻ Integration of Furniture, Fixtures, and Equipment

132. B
133. B
134. A
135. C
136. B
137. C
138. A, B, C, F
139. B, C, D, E
140. A, B, D, F
141. D
142. B
143. A, B, C, E
144. C
145. D
146. A
147. B
148. A, B, D, F
149. A, B, E, F

❼ Contract Administration

150. A
151. A
152. D
153. C
154. B
155. A
156. B, C, D, E
157. A, B, D
158. C
159. C, D, E, F
160. D
161. D
162. A, B, D, E
163. A
164. A, C, D, E
165. B, D, E
166. A, D, E, F
167. B
168. C
169. C
170. A, B, E, F
171. D
172. C
173. C
174. B
175. A

❶ PROJECT ASSESSMENT AND SUSTAINABILITY

1. ***The answer is B.*** The building load factor accounts for each tenant's share of the common areas of the building, such as restrooms, public corridors, and elevator lobbies. The building owner calculates the load factor using one of two methods, Method A and Method B, as defined by the ANSI/BOMA Z65.1 standard. Once the gross floor area of a tenant space is known, it is multiplied by the load factor to determine the total square footage upon which the rent calculations will be based.

The building efficiency ratio is the ratio of the net assignable area of a building to the gross building area. The R/U ratio is the rentable area divided by the usable area of each floor in an office building and is used by Method A of ANSI/BOMA Z65.1 to determine the load factor, which may vary from floor to floor. Although basically the same as the load factor referenced in Method A, the correct term is *building load factor* because Method B of ANSI/BOMA Z65.1 is used to develop a building load factor that is the same for the entire building. The rentable area is what the tenant pays rent on.

2. ***The answer is A.*** Daylighting zones are areas near windows, roof monitors, and skylights where daylight-responsive controls are required by the *International Energy Conservation Code* (IECC). The IECC defines where these zones are located based on the size of the fenestrations, ceiling height, and obstructions. In section view, the daylight zone extends as far into the building from the window as the distance from the floor to the top of the window. In plan view, it extends 2 ft (610 mm) on either side of the window jamb, or to the nearest wall, whichever is less. When planning the interior configuration, the designer should determine where these spaces are and locate uses within them that can function well with daylight-responsive controls. Other types of lighting controls should be specified for spaces beyond these zones.

HVAC zones are established by the mechanical engineer and generally function independently of areas that the interior designer may have control over, which might amount to any significant energy savings. Occupant sensor controls, although important for energy savings, only detect the presence or absence of people within an area and turn the lighting or other devices on or off. They do not relate directly to space planning, especially during design development. Time-switch controls turn off lights or other loads automatically based on time schedules, such as overnight or on holidays.

3. ***The answer is C.*** The contractor is the person typically responsible for submitting the plans and specifications to the building department, along with the application for a permit.

4. ***The answer is A.*** Although all the options given contain important parts of a due diligence investigation, the basic definition is a thorough understanding and documentation of the space in which a client's project will be located, as well as its surrounding context.

5. ***The answer is B, C, E, and F.*** The type of community, the amount of tenant parking, adequate bicycle facilities, and the location of public transportation are all factors that are relevant to LEED-CI certification. The type of streets leading to the project site has no relevance on a building's ability to achieve LEED-CI certification. Access to sunlight is not addressed in the possible LEED-CI credits.

6. ***The answer is B.*** A nurses' workstation requires much record keeping and the temporary and long-term storage of a variety of items. The work surface and storage requirements will include provisions for most of the elements that use large amounts of space, including report writing and the number of patient files and moveable carts. Electrical and communication equipment requirements do not contribute significantly to the space needed for the nurses' station.

7. ***The answer is D.*** An interior designer would need to consult a structural engineer to determine the feasibility of modifying a loadbearing wall. The information listed in options A, B, and C could easily be determined through simple inspection by someone on the designer's team. The adequacy of the air supply would have to be determined by a mechanical consultant, but the number of diffusers could easily be counted. Water pressure for one additional sink could be verified by turning on a faucet.

8. ***The answer is B.*** To determine if a site and its potential use are compatible, an interior designer and a client must first develop a document including a program, a list of spaces needed in the building and their approximate sizes. This document will serve as the "instruction manual" for the project. The interior designer in this scenario should work with the client to develop this document and compare it against the existing space to see if the new use is compatible with the old structure. If it is determined to be a good fit, the process of site analysis can move forward. After that it will be important to study code issues, parking requirements, environmental concerns, and the historic value of the property to determine if the project is feasible. Simultaneously, the client should be developing the project pro forma, which examines the financial feasibility of the project. If the results of both studies indicate that the project should go ahead, the interior designer may begin to prepare schematic design concepts for the renovation.

9. ***The answer is B.*** According to BOMA standards, when 50% or more of the exterior wall is comprised of glass, measurement is taken to the inside face of the glazing.

10. ***The answer is B, D, E, and F.*** The building common area means the areas of the building that provide services to building tenants but which are not included in the office area of any specific tenant. These areas may include such areas as lobbies, public corridors, lounges, vending areas, health centers, and service areas such as fully-enclosed mechanical or equipment rooms.

11. The load factor accounts for each tenant's share of the common areas, including the public corridor, restrooms, mechanical rooms, and elevator lobby. To determine the rentable area, the load factor is multiplied by the tenant's gross occupied area. In this case

$$\text{rentable area}_{\text{U.S.}} = \left(4500\ \text{ft}^2\right)(1.15) = 5175\ \text{ft}^2$$

$$\text{rentable area}_{\text{SI}} = \left(418\ \text{m}^2\right)(1.15) = 480.5\ \text{m}^2 \quad \left(481\ \text{m}^2\right)$$

12. ***The answer is B, C, E, and F.*** WaterSense is a program of the Environmental Protection Agency (EPA) that helps consumers identify water-efficient programs and products, which are certified by independent, third-party bodies based on EPA's criteria for water efficiency. High-efficiency pre-rinse spray valves can minimize water use as well as drainage runoff. Although not directly under the interior designer's control, a post-occupancy suggestion to the client to develop an ongoing program to find water leaks can save water over the long term. On average, leaks can account for more than 6% of a facility's total water use. As with WaterSense, using ENERGY STAR appliances can minimize water use. ENERGY STAR-qualified commercial dishwashers are about 50% more water efficient than standard models. Icemakers are from 16% to 20% more water efficient than standard models.

The problem statement does not make it clear if the restaurant is freestanding or inside an existing building, so any decisions on graywater use cannot be made. In addition, the *Uniform Plumbing Code* (UPC) does not allow graywater to be used inside a building unless it is treated with an on-site water reuse system.

Reducing the temperature of the water would not have any effect on the actual quantity of water used, only on the energy required to heat it. Hot water is defined by the *International Plumbing Code* (IPC) as being water at a temperature of 110°F (43°C) or greater. However, local or state regulations generally require a water temperature ranging from 165°F (74°C) to 180°F (82°C) for sanitizing dishes.

13. ***The answer is A, B, D, and E.*** The basic causes of poor indoor air quality (IAQ) are contaminants from outdoor and indoor sources, biological contaminants, and poor ventilation. Indoor sources may include things such as volatile organic compounds (VOCs), tobacco smoke, and carbon monoxide. Biological contaminants may include mold, viruses, and bacteria. Poor ventilation allows indoor pollutants to accumulate to unpleasant or even unhealthy levels.

Poor building maintenance is generally not a major problem for poor indoor quality unless it results in not removing biological contaminants or cleaning chemicals. Sick-building syndrome is not a source of poor indoor air quality. It is a condition that cannot be directly linked to any particular cause.

14. ***The answer is D.*** Displacement ventilation is an air distribution system in which supply air originates at floor level through registers and ductwork below a raised floor and rises to return-air grilles in the ceiling. The supply air can be delivered close to users and equipment so the air does not have to be cooled as much, resulting in energy savings. The raised floor could be used to provide electrical and data service to the computers as well as provide space for ductwork. An all-air system would work but would not be as efficient as displaced ventilation. A low ceiling could also make it difficult to provide ductwork in the ceiling plenum. The project would already be subject to the *International Energy Conservation Code* (IECC) and meet the requirements in that code, which could be met by any number of mechanical systems. While placing the computer area away from windows would make its conditioning easier, displacement ventilation would most likely be used in conjunction with space planning.

15. ***The answer is C.*** The removal of vermiculite requires the services of a trained and certified professional contractor. Formaldehyde is inherent in many building products, so it really cannot be removed without removing the product it is a part of. It is best mitigated by the interior designer specifying materials that do not use formaldehyde or that include it in acceptable amounts. If commercial cleaning compounds contain an unacceptable level of chemical contaminants, they can easily be removed and properly disposed of. As with formaldehyde, volatile organic compounds are within a product so the interior design or architect should specify material that have no volatile organic compounds (VOCs) or those that meet acceptable limits.

16. ***The answer is B, C, and D.*** One of the primary goals of retail design is to encourage sales, and therefore profit, through design to encourage people to shop and make purchases. For this reason, the developer, store owner, and the public would all have significant influence on how the designer approached the work.

The leasing agent would be interested mainly in keeping the space rented. While the employees may appreciate working in an attractive space, they would be most interested in having a job. The architect would be more concerned with the overall design of the complex and the core and shell of the department store.

17. ***The answer is A.*** Linoleum is made from renewable products, including linseed oil, rosin, cork powder, and pigments. Carpet made with polyethylene terephthalate (PET), while a good sustainable choice because it is made from recycled soft drink containers, is not itself a renewable material. Rubber is also a recycled material but not renewable. Vinyl is made from highly refined petrochemicals and is not a renewable option.

18. ***The answer is 10,000 ft² (900 m²). (Building A has the smaller rentable area.)***

Customary U.S. Solution

First, calculate the occupant area to account for required circulation space. For building A, the estimated efficiency is 80% or 0.80.

$$\begin{aligned}\text{occupant area}_{\text{A}} &= \frac{\text{net assignable area}}{0.80} = \frac{6500\ \text{ft}^2}{0.80}\\ &= 8125\ \text{ft}^2\end{aligned}$$

Next, calculate the rentable area for building A. Using the leasing agent's building load factor of 1.23, multiply the occupant area by 1.23.

$$\begin{aligned}\text{rentable area}_{\text{A}} &= (\text{occupant area}_{\text{A}})(1.23) = (8125\ \text{ft}^2)(1.23) \\ &= 9994\ \text{ft}^2\end{aligned}$$

For building B, the estimated efficiency factor is 75% or 0.75.

$$\begin{aligned}\text{occupant area}_{\text{B}} &= \frac{\text{net assignable area}}{0.75} = \frac{6500\ \text{ft}^2}{0.75} \\ &= 8667\ \text{ft}^2\end{aligned}$$

Calculate the rentable area for building B.

$$\begin{aligned}\text{rentable area}_{\text{B}} &= (\text{occupant area}_{\text{B}})(1.20) = (8667\ \text{ft}^2)(1.20) \\ &= 10{,}400\ \text{ft}^2\end{aligned}$$

Of the two buildings, building A has the smaller rentable area, 9994 ft^2 (10,000 ft^2).

SI Solution

First, calculate the occupant area to account for required circulation space. For building A, the estimated efficiency is 80% or 0.80.

$$\begin{aligned}\text{occupant area}_{\text{A}} &= \frac{\text{net assignable area}}{0.80} = \frac{604\ \text{m}^2}{0.80} \\ &= 755\ \text{m}^2\end{aligned}$$

Next, calculate the rentable area for building A. Using the leasing agent's building load factor of 1.23, multiply the occupant area by 1.23.

$$\begin{aligned}\text{rentable area}_{\text{A}} &= (\text{occupant area}_{\text{A}})(1.23) = (755\ \text{m}^2)(1.23) \\ &= 929\ \text{m}^2\end{aligned}$$

For building B, the estimated efficiency factor is 75% or 0.75.

$$\begin{aligned}\text{occupant area}_{\text{B}} &= \frac{\text{net assignable area}}{0.75} = \frac{604\ \text{m}^2}{0.75} \\ &= 805\ \text{m}^2\end{aligned}$$

Calculate the rentable area for building B.

$$\begin{aligned}\text{rentable area}_B &= (\text{occupant area}_B)(1.20) = (805\ \text{m}^2)(1.20) \\ &= 966\ \text{m}^2\end{aligned}$$

Of the two buildings, building A has the smaller rentable area, 929 m^2 (900 m^2).

19. ***The answer is A, C, D, and F.*** The interior designer has control to select low volatile organic compound (VOC) materials, space plan to isolate high polluting areas, specify building commissioning, and verify, either with the architect or with the mechanical engineer, if adequate outdoor ventilation is being provided. Although the designer can suggest a no-smoking policy and that the client monitor spaces after occupancy, they cannot control what happens after occupancy.

20. ***The answer is D.*** Many doorways at first viewing look acceptable and may be overlooked. It is only with measurement that the designer can determine if the doorways meet accessibility requirements. Unacceptable floor level changes and protruding objects are obvious. Ramps are always a red flag for accessibility compliance and are likely to be carefully checked.

21. ***The answer is B.*** Knowing the seismic category will tell the designer if special ceiling details need to be specified. This should be the first course of action. If the project happens to be in category A or B, no special details are required. Part of determining the seismic category of a building depends on the soil under the building, so the categories may vary from one area to another even within the same jurisdiction. The results of soil tests will be used by a structural engineer to make the correct determination. There would be no need to include seismic details if they are not required. The ceiling contractor would not take on the responsibility of seismic details without the direction of the interior designer.

22. ***The answer is B.*** Cost evaluations performed during the programming stage often compare a proposed project to a model project of similar size and scope. A premium is something that will add cost to a project in comparison to the model. Examples of premiums are short construction periods, unusual contract provisions (extra insurance, liquidated damages, etc.), and nonstandard programmatic elements or client requirements, such as the need to use prevailing wage rates or union labor.

A contingency is just an amount set aside for unknown circumstances. An add-alternate is an item for which the contractor must give a price during bidding. An upcharge is an extra amount the contractor may apply for something, which is unknown to the interior designer.

23. ***The answer is C.*** Goals and objectives are overriding, general desires on the part of the client. They are the basis for the interior designer to develop programmatic concepts, which then lead to design concepts, which ultimately lead to physical solutions to meet the client's goals and objectives. Goals and objectives precede how the client wants to organize the project, which groups of users are of primary importance (although they may be part of the objectives), and specific space requirements.

24. ***The answer is A.*** A change in use may result in a change in occupant load, which could exceed the allowable occupancy for a given area. The building code should be consulted for the maximum allowable area, which is based on occupancy group, type of construction, and whether the building is sprinklered or not. For example, changing from a business occupancy (150 ft^2 per occupant, gross) to a restaurant assembly occupancy (15 ft^2 per occupant) may make a project infeasible or require fire walls to separate the space into smaller fire areas.

A change in use may affect the number and location of exits but this determination is made after verifying the allowable area of the new use. Comparing maximum travel distances is not relevant. Only the travel distance of the final space plan would be important. Determining the existing construction type, while necessary, is not sufficient without also determining if fire suppression is present and reviewing tables in the *International Building Code* (IBC), which give the maximum allowable areas.

25. ***The answer is A, B, C, and D.*** The interior designer has the ability to make sure CFC-based refrigerants are not used by the mechanical engineers, to specify automatic lighting controls or instruct the electrical engineer to do so, to easily plan reading rooms to take advantage of daylighting, and to design or specify automatic shading systems to control glare and encourage daylighting.

While book stacks could moderate glare it would not be a wise use of available daylighting. Displacement ventilation could be used but this would require an extensive and expensive installation of a raised flooring system. Energy efficient HVAC could be accomplished with other less expensive methods.

26. ***The answer is D.*** Locating the development on a brownfield site can receive points in the LEED rating process. A brownfield site is commercial or industrial property where the development, expansion, or reuse may be complicated by the presence or potential presence of a hazardous substance, pollutant, or contaminant. Insulating glass would help limit heat loss but would not, in itself, achieve LEED credits. It is unlikely that existing partitions would serve to make it easier to build out a retail space, so reuse would have negligible use for LEED credits. Parking must meet minimum zoning requirements and have a minimum of 50% underground or covered to receive LEED credits.

❷ PROJECT PROCESS, ROLES, AND COORDINATION

27. ***The answer is C.*** Although all of these professionals may need to be consulted regarding book stacks, option C is the best choice. Because book stacks in libraries are very heavy, a structural engineer needs to determine if the existing floor is capable of supporting the weight. This would be the most important determining factor in early space planning and stack location. Issues of air supply, stack types, and sprinkler locations could be based on the final locations of stacks.

28. ***The answer is A, C, E, and F.*** New window coverings can affect the exterior appearance of a building, which might be objectionable to the building owner. They can also affect the mechanical system, so the mechanical engineer should be consulted. The new coverings must meet flammability requirements, so the manufacturer should be consulted. Also, the location of the window covering relative to the glass can put additional heat stress on the glass, causing cracking or breaking.

Although light reflectance might be affected, it would be minor and probably would not affect the overall light quality in the room. While the client's opinion about color is important, it is not the most critical coordination item, because once the other factors have been coordinated, an acceptable color and pattern can found.

29. ***The answer is D.*** The interior designer would most likely be able to influence the design of sprinklers by coordinating with the mechanical or fire protection engineer during the construction documents phase of a project. For example, the interior designer could request that sprinklers be placed in certain positions or that additional sprinklers be installed. Compartmentation and smoke control are already designed by the architect and mechanical engineer in the original plans of the building, and there would be little the designer could do to change these elements. Fire detection is determined by the local building code and the type of building and occupancy, so little influence by the designer is possible.

30. ***The answer is B.*** Although the interior designer will indicate the desired location of telephone and data outlets based on client feedback obtained during the programming process, the electrical engineer is responsible for showing the actual conduit runs for the wiring.

31. ***The answer is A.*** The custom-designed reception desk section includes an undercounter light and an electrical connection that is hard-wired through the floor. Therefore, the electrical engineer needs to be consulted to confirm the location of electrical stubs on the electrical plans, and to confirm the requirements for conduit size, access panels, and types of connection.

32. ***The answer is D.*** For custom-designed woodwork, the woodwork fabricator orders minor accessory materials such as stone, metal, and glass from other suppliers, and then incorporates them into the woodwork.

33. ***The answer is C.*** The interior designer is ultimately responsible for coordinating the drawings of the various consultants.

34. ***The answer is C.*** The owner is ultimately responsible for deciding which contractor to award the contract to. The interior designer is generally involved but only assists with the process and gives advice.

35. ***The answer is B.*** In this situation, the dealership would most likely provide the ordering, delivery, and installation services, including the paperwork needed.

36. ***The answer is D.*** The location and general condition of existing light fixtures requires no electrical expertise, and these can be verified by the interior designer. However, such things as the capacity of wiring, the condition of wiring, and the safety of the fixtures must be determined by the electrical engineer or a qualified electrician.

37. ***The answer is B.*** The critical path method (CPM) uses a network diagram (i.e., chart) that shows all tasks required to complete a project. From the CPM chart, a path of the critical tasks defines the sequence of tasks that must be started and finished exactly on time for the schedule to be met. CPM charts also can include an analysis of the time/cost tradeoffs. The program evaluation review technique (commonly referred to as a PERT chart) is similar to the CPM method, but uses different networking methods. PERT charts are not commonly used with complex projects because CPM charts better show critical tasks and their associated times. A bar chart (also called a Gantt chart) is a simpler representation of a schedule and does not show a critical path.

38. ***The answer is D.*** A full wall schedule is interactive and involves the entire project team—including the client—in the project scheduling. Each person on the project team places a card with time estimates for their associated tasks on the schedule. Then, the schedule serves as a starting point for discussion, as the cards can be moved around until everyone agrees on the project schedule. Once a schedule is set, it can be copied in smaller format for each project team member.

39. ***The answer is A, B, D, and F.*** The interior designer is responsible for giving the electrical and mechanical engineers information about equipment or occupants so that they can accurately calculate electrical loading and HVAC system requirements. The interior designer also has overall responsibility for making sure all the consultants' documents are coordinated with the interior design documents.

The mechanical engineer (or fire protection engineer, if there is a separate consultant) is the person responsible for staying current with all code requirements in the project's jurisdiction. The location of the electrical closets is predetermined by the building architect, not the interior designer.

40. ***The answer is C.*** Lighting locations are shown on the interior designer's reflected ceiling plans. All the other options listed would be on the electrical engineer's drawings or on the security consultant's drawings.

41. ***The answer is D.*** In addition to designing the HVAC system, the mechanical engineer determines the best locations for controls such as thermostats. When these locations interfere with other wall-mounted items, the interior designer may sometimes suggest minor modifications.

42. ***The answer is B.*** In the sequence of construction, the mechanical ductwork and other mechanical equipment are the first things to be installed, and also the most difficult to modify once in place. The interior designer should discuss and verify with the general contractor and mechanical subcontractor what areas of the space will need sufficient room to provide for luminaires and raised ceilings.

Electrical work is usually the last service to be installed and can easily be routed to avoid recessed lighting and raised ceilings. Plumbing is typically not extensive, is installed after the mechanical work, and can usually accommodate the location of other services. Sprinkler piping is typically installed after the mechanical work and provides some flexibility in avoiding recessed lighting and raised ceilings.

43. ***The answer is D.*** A line chart lists the various products a representative handles.

44. ***The answer is D.*** For a product like acoustic insulation, a number of manufacturers' products will perform equally well, as long as the material possesses the minimum characteristics defined in the specification, including acoustical performance, flammability, thickness, and UL compliance when used in fire-rated applications. A reference specification would give the contractor the latitude to request bids from various suppliers to get the lowest price, as long as the proposed products meet the reference standards.

A performance specification for this type of product would be unnecessary because many standard products could meet the requirements. A prescriptive specification tells the contractor exactly which product or manufacturer to use. A proprietary specification is a type of prescriptive specification that calls out a specific brand name for a product and would not be necessary for acoustic insulation.

45. ***The answer is A.*** Because general contractors add their overhead and profit charges (anywhere from about 10% to 20%) to all subcontracted work, the client would be paying that much extra for the appliances without the general contractor doing much work for the extra cost. Option D is incorrect because the interior designer could get about the same discount for the client as the contractor could, without the contractor's markup.

46. ***The answer is A, B, C, and E.*** The owner is required to provide a written program, laboratory tests specified by the contract documents, removal of existing furniture, and space for materials used on the project according to AIA Document B152. Although the owner ultimately pays for copies of drawings as a reimbursable expense, the responsibility for providing them to the contractor rests with the interior designer. The owner is involved with the bidding process, but the interior designer is responsible for preparing the bidding documents.

47. ***The answer is B, C, and D.*** Reviewing laws and regulations, developing a schedule of decision dates, and maintaining a record of changes are all part of the standard AIA Document B152 agreement. B152 states that inspecting and accepting furniture, conducting an existing furniture, fixtures, and equipment (FF&E) inventory, and appearing before legal or public proceedings are additional services of the interior designer, and are only undertaken if those activities are specifically made part of the agreement.

48. ***The answer is A, D, E, and F.*** A negotiated contract is established when a client, with the assistance of the interior designer, chooses a contractor (and, if desired, the subcontractors) based on factors other than cost. Once the choice has been made, the client and contractor negotiate the construction cost and other aspects of the contract.

The process of establishing a bid contract begins when an interior designer prepares and sends project drawings and specifications to several general contractors. The contractors then bid on the work, and the client chooses a contractor by evaluating the bids and other relevant factors. This type of contract generally results in the lowest contract price but the longest project schedule, because of the protracted bidding process. Although selecting a contractor using many criteria usually results in a quality project, it is not guaranteed.

49. ***The answer is A, B, E, and F.*** The evaluation of building systems, materials, and operation are always part of a post-occupancy evaluation. This includes talking to the users, evaluating possible maintenance problems, and evaluating contractor performance. Although the interior designer may use lessons learned from one job to the next, suggestions for improvement are generally not placed in a verbal or written post-occupancy evaluation. Construction cost review with the client is also not a normal part of post-occupancy evaluation, although it may provide valuable information for the interior designer in estimating future projects.

50. ***The answer is B.*** Project tracking involves comparing how a job is progressing in a design office, both during design and construction, compared with the original estimated time and fee projections. Typically done on a weekly basis, the review of time and fees gives the project manager the ability to make modifications as necessary to meet the original financial goals of the project.

While using time sheets provides raw data for tracking a project, using them to measure productivity and billing rates is not a type of project tracking. Comparing client goals during construction is a futile effort as the project is already underway and is something that should be done during schematic design and design development phases. While keeping track of actual construction costs against the original construction cost estimate is vital, it is not project monitoring.

51. ***The answer is B.*** The base bid with "approved equal" language is the best in this situation. It lists the desired product and states that an alternative product proposed by the contractor will be considered by the interior designer. This establishes a minimum level of quality based upon the characteristics of the specified product, but it puts the responsibility on the contractor to find and submit an alternative product if they wish. This type of specification will require the interior designer to evaluate the proposals during the bidding phase and issue addenda notifying all bidders of the decisions if an alternate is approved.

A proprietary specification is not the best choice in a situation where the owner and interior designer wish to encourage competitive bids. The proprietary specification specifies a particular product by brand name and allows no substitutions. A descriptive specification defines the type of outcome desired but does not list specific products. The descriptive specification is the most difficult type of specification for an interior designer to write because it requires listing all of the criteria that a material or assembly must meet. A base bid with alternate specifications is similar in format to the base bid with "approved equal" language specification. Both call for a specific product but allow substitution of other materials. However, an important difference between the two is that the base bid with alternates allows a contractor to substitute a product that they feel is equal and does not require the interior designer's approval. The product submitted may not be comparable to the one defined in the specification but must be accepted due to the way the specification is written.

52. ***The answer is B, D, and F.*** One of the primary goals of quality control through design, as well as during subsequent phases of a project, is to meet the client's original, stated goals and objectives as documented in the program. These can range from the obvious ones like keeping costs to a minimum to more lofty goals like creating the best place for employees to work. During this time, the interior designer must also provide the expected standard of care and reduce the designer's own risk and liability. All of these should be achieved by using the available employees working on the project. Designing an award-winning project may be a goal of the designer but will be of little importance if the final project turns out costing too much or not meeting the client's expectations. While the project may or may not meet the expectations of the future users, those are undefinable goals the designer cannot anticipate unless they are made part of the client's original goals.

❸ PROFESSIONAL BUSINESS PRACTICES

53. ***The answer is A.*** Practice acts require certain minimum educational requirements as well as passing the National Council for Interior Design Qualification (NCIDQ) exam. Not all states require continuing education to maintain the use of a title. The allowable titles in states with title acts use different titles such as "certified interior designer," "registered interior designer," or "interior designer." Although any designer who has passed the NCIDQ exam would probably not want to call themselves an interior decorator, practice acts do not prohibit it.

54. ***The answer is D.*** The client must sign the sales agreement, which obligates the client to pay for the purchase of merchandise. The other documents are handled by the interior designer, dealership, or others.

55. ***The answer is A.*** Aged accounts receivable are accounts with invoices that are still unpaid after a certain length of time, such as 90 days. A list of aged accounts receivable should be kept and regularly updated, and used to follow up with clients who have outstanding invoices.

56. ***The answer is B.*** Cash accounting is the simplest type of accounting. It is appropriate for a self-employed person as well as for a very small firm, as long as the firm is not a corporation, which cannot use cash accounting.

57. ***The answer is D.*** A quantity takeoff is the most detailed method of developing a budget, and therefore it is the most accurate. To use this method, count the actual quantities of materials and furnishings and multiply these quantities by firm, quoted costs.

58. ***The answer is C.*** Net price is the final cost after all discounts and any other price adjustments are done by the furniture dealership. The manufacturer's price list or a showroom manager could provide a price for the furniture, but the price would not include discounts or other price adjustments. Therefore, the interior designer's best option for determining the net price is the furniture dealership. Line charts have nothing to do with prices.

59. ***The answer is D.*** Under the AIA Document B152, the interior designer is required to develop a preliminary cost estimate (based on area, volume, or a similar conceptual estimating technique) and update it at the schematic design, design development, and construction documents phases.

60. ***The answer is D.*** A labor and material payment bond is designed to pay subcontractors and vendors in case the general contractor defaults on his or her payments for labor and materials provided. When subcontractors or vendors are not paid for their work, they can file liens against the property. A performance bond provides money for completion of a project should the general contractor default, but it does not provide for payment of past-due bills on the original construction.

61. ***The answer is B.*** Because bond money is a fixed amount, the budget must be met, so this eliminates option A. Rebidding takes additional time and does not guarantee that the new bids will be any better; in fact, they may be higher because prices will probably increase in the time it takes to rebid. This eliminates option C. The designer may want to wait for direction from the city, but the project must go forward. The amounts of the bids are so close to the budget that it is likely that costs could be reduced by 4% with some adjustments in the scope of the project.

62. ***The answer is C.*** For bidding, the procedure a contractor must follow to propose a substitution is in the instructions to bidders. After the contract is awarded, the specification requirements can be found in the General Requirements. The advertisement to bid simply states that bidding is being accepted for a particular project and gives information about how to submit a bid. There is no such document as "bidding procedures."

63. ***The answer is A.*** An allowance compensates for costs that are unknown at the time of a bid. An add-alternate requires contractors to provide alternate prices for something that varies from the base bid. A bonding company guarantees payment for materials using a material bond. Contractors generally commit to a price on a portion of work before the total quantity of the work is known using a unit price.

64. ***The answer is C.*** Because the two lowest bids are so close, it is likely that they represent a true indication of the cost for the restaurant as designed rather than an overbid. Although the client has the option of trying to get more money, it is generally the designer's responsibility to be within 10% of the expected bid. For this reason, the designer should offer to help the client redesign as necessary to reduce the cost.

65. ***The answer is A, B, C, and F.*** Even though a sculpture is physically attached to the construction, a work of art would be commissioned directly with the artist and included in the furniture, fixtures, and equipment (FF&E) budget. Vending machines can also be part of a construction contract, but if there is both a construction contract and an FF&E contract, they are either part of the FF&E contract or leased directly from the vending company. Auditorium seating is attached to the structure and considered "furnishings" in CSI MasterFormat Division 12. Vertical blinds are sometimes included in the construction contract if they are motorized or require special construction detailing, but they are not as commonly included in the construction contract as wall-to-wall carpeting is.

Carpeting and applied acoustical panels are finish items that are attached to the construction and are typically included in the construction contract, not the FF&E contract.

66. ***The answer is B, C, D, and E.*** Estimates for built-in equipment, fixtures to be supplied by the owner, contractor's profit, and custom chandeliers installed by the contractor all relate to the actual construction of the project, including items that may be purchased separately by the owner but installed by the contractor. Designer's fees and allowances for area rugs—which are considered furniture, fixtures, and equipment (FF&E)—are not included in a construction budget.

67. ***The answer is B.*** During bidding, a unit price is a set price quote from a contractor for a portion of the work based on individual units of measurement, such as a price per square foot or per linear foot (per square meter or per linear meter). A unit price is often requested in addition to a base bid when the full extent of work is not yet known to help compare the costs of different contractors for the same work.

68. ***The answer is D.*** The *Uniform Commercial Code* (UCC) assigns risks by allowing the factory and vendor to use "FOB factory" or "FOB destination" to determine at what point title is transferred and who is at risk for shipping damage.

69. ***The answer is D.*** Under AIA Document A251, the owner is responsible for conducting an acceptance inspection, with the *assistance* of the architect or interior designer.

70. ***The answer is B.*** In most jurisdictions, a business that buys merchandise at wholesale and resells it is required to have a transaction privilege tax license, also called a resale license or sales tax license in some states.

A business license of some type is usually all that is required to conduct business. No further license is required to charge clients for the cost of employees or for installation services.

❹ CODE REQUIREMENTS, LAWS, STANDARDS, AND REGULATIONS

71. ***The answer is B.*** Scoping provisions tell the designer how much of something is required. Although there are scoping provisions in the ADA, local codes may be more stringent, in which case the interior designer must conform to the more restrictive requirements.

ANSI A117.1 may or may not be applicable in a given jurisdiction. If ANSI A117.1 has been adopted by the local building code, it should be reviewed to see if the requirements are more restrictive than those of the ADA. The latest edition of A117.1 leaves scoping up to local jurisdiction. The *ADA/ABA Guidelines* are the *ADA/ABA Accessibility Guidelines* and are part of the ADA, so this option is irrelevant. The *Uniform Federal Accessibility Standards* are applicable to federal buildings and projects that receive federal funding and would not be used for a commercial project.

72. ***The answer is D.*** Any rights not specifically reserved for the federal government by the U.S. Constitution revert to individual states. The states, in turn, can delegate control of construction to local jurisdictions. Only a few states have a state building code. In nearly all cases, the local or state code is based on the *International Building Code* (IBC) or one of the model codes.

73. ***The answer is B.*** ASTM International, formerly known as the American Society for Testing and Materials, is one of the organizations that establish a wide variety of standards covering testing methods, products, definitions, and more. Although its committees develop test methods, it is not a testing laboratory.

74. ***The answer is D.*** Using the space-by-space method, an LPD in W/ft^2 is assigned by the IECC for various types of spaces. This value is multiplied by the area of the space under consideration to arrive at the total allowable power that can be used for lighting that space. Knowing this value will assist the designer in selecting lamp and luminaire types to meet the illumination requirements of the space while keeping within the power budget. Because there are many individual space types given in the IECC, the designer would most likely use this method to meet the specific illumination requirements of each space. Further, this method allows for tradeoffs, so if less power is required or used for one space, the extra power unused in one area can be transferred to the allowance for another space.

The building area method for developing lighting power densities (LPDs) uses an LPD for a general building type, which is then multiplied by the total area of the building to get the total allowable power that can be used for the building. It would probably not be used by an interior designer to design just a portion of a larger building. The energy cost budget method is a more complex way to determine the total energy budget for an entire building using computer simulation of hourly energy use over the course of a year. It is not appropriate for the interior design of specific spaces. The luminaire efficacy method is incorrect because there is no such method.

75. ***The answer is C.*** The Steiner tunnel test (ASTM E84) is used to measure the flammability of wall finishes and is the test most often required in building codes. The methenamine pill test is for carpet. The smoke density test does not measure flame spread, which is of vital importance. The vertical ignition test is for window coverings.

76. ***The answer is D.*** The LEED rating systems are green building certification guidelines developed by the U.S. Green Building Council (USGBC). There are five levels of certification (effort, certified, silver, gold, and platinum) set forth in the Green Business Certification Inc. (GBCI, formerly the Green Building Certification Institute) rating systems.

Options A and B are incorrect because Greenguard and Green Seal are both product-rating systems. Greenguard certifies for acceptable emission levels, while Green Seal certifies products that meet certain environmental standards. Option C is incorrect because ISO 14000 refers to the International Standards Organization's collection of standards and guidelines that relate to a variety of environmental standards, including labeling, life-cycle assessment, and others. ISO standards are used to measure the performance of other organizations that certify products' other environmental claims.

77. ***The answer is D.*** Remodeling the building from a residence to a restaurant involves a change in use, which is regulated by the zoning department. If the mansion is in a zoning district that does not allow business uses, the project would either not be feasible or require that the owner apply for a variance. Neither the building department nor the city planning commission regulates changes in use. The department of excise is concerned with tax issues, not building issues.

78. ***The answer is D.*** By definition, a material that does not ignite or burn is considered noncombustible.

79. ***The answer is A.*** Standards for VOC limits on furniture systems are established by BIFMA e3, *Furniture Sustainability Standard.* It is applicable to movable walls, systems furniture, desking systems, casegoods, tables, seating, and accessories. Annex C of the standard lists concentration limits for over 30 individual VOC chemical compounds for workstations, seating, and individual components. The e3 standard references ANSI/BIFMA M7.1, *Standard Test Method for Determining VOC Emissions from Office Furniture Systems, Components and Seating,* and ANSI/BIFMA X7.1, *Standard for Formaldehyde & TVOC Emissions of Low-emitting Office Furniture Systems and Seating.*

The Environmental Protection Agency (EPA) establishes limits only on VOCs for coatings. The *International Building Code* (IBC) and *International Energy Conservation Code* (IECC) do not establish any limits on VOCs.

80. ***The answer is B.*** A life-cycle assessment evaluates the environmental impact of using a particular material over its entire useful life, including disposal. It could be used to compare the impacts of two or more materials so the architect could select the most sustainable one.

Option A is incorrect because an environmental impact study is used to evaluate the impact of a development on the environment. Option C is incorrect because an impact assessment is one phase of a life-cycle assessment. Option D is incorrect because there is no sustainability evaluation method by that name.

81. ***The answer is D.*** If a product has been tested according to ASTM standards or otherwise meets ASTM requirements, then the manufacturer should have the test results or other backup information and be willing to send that information to the designer on request.

82. ***The answer is A.*** ASTM D4157 is the Wyzenbeek abrasion resistance test and is used only for fabrics, not for the construction of the chair itself. CAL TB133 is the full seating test and is used for flammability testing. NFPA 701 is the vertical ignition test for the flammability of draperies or window treatments.

83. ***The answer is C.*** The IECC, *International Energy Conservation Code*, is a model code that regulates minimum energy conservation requirements for both residential and commercial buildings for all aspects of energy use. *The International Green Construction Code*, or IgCC, also contains requirements for energy conservation but is oriented more for engineers and also incorporates other technical codes. The IBC, *International Building Code*, references the IECC but doesn't provide as much information and prescriptive criteria as the IECC. The IRC is the *International Residential Code* and applies only to residential buildings.

84. ***The answer is B.*** When the interior designer completes the construction documents, they are given to the contractor who is responsible for submitting the documents to the authority having jurisdiction along with a permit application form and the necessary fee.

85. ***The answer is D.*** Zoning ordinances determine allowable uses depending on the location in a city or jurisdiction. When a change in use is proposed for a building, the interior designer must verify if the change is allowable. If not, the client and designer may elect to apply for a variance to allow the non-conforming use.

The building department would be mainly interested in things such as occupancy limits of the new use, means of egress, fire protection, and the like. A planning commission and city council would only become involved if the proposed change in use violated a master plan of the city and a change in zoning was required.

86. ***The answer is B.*** The Fair Housing Act (and state laws) governs accessibility requirements of multifamily housing. While there are many similarities to other Americans with Disabilities Act (ADA) standards, there are some unique features. The *International Building Code* (IBC) and the *Uniform Federal Accessibility Standards* do not cover all aspects of multifamily housing.

87. ***The answer is C.*** A common path of egress travel is the part of the path of egress travel that occupants are required to move through before they reach a point where they have a choice to follow one of two distinct paths of travel to an exit.

88. ***The answer is B, C, and D.*** By definition, the means of egress consists of the exit access, the exit, and the exit discharge.

The means of egress must lead to a public way, but the public way is not a part of the means of egress. The common path of travel is that portion of exit access that the occupants are required to traverse before two separate and distinct paths of egress travel to two exits are available. Although an important feature, this is not a part of the means of egress according to model codes. The travel distance is the distance from any point in the exit access to the nearest exit.

89. ***The answer is C.*** The lighting power density (LPD) is a value that the *International Energy Conservation Code* (IECC) assigns to various space types. The LPD is the maximum amount of power per square foot that can be used in a space for lighting. The value is multiplied by the area of the space to arrive at a total amount of power (wattage) that can be apportioned to the luminaires. This is known as the lighting power allowance (LPA) and is used in the space-by-space method of calculating the total maximum interior lighting allowance for a building or area in a building.

The energy cost budget method considers total energy use in a building, not just lighting. The designer can then make trade-offs between lighting and other energy needs. It is a complex method using computer simulation compared to a baseline building. The space-by-space method uses the LPA for individual spaces to determine the lighting power density and to arrive at the total allowable energy use that can be applied to lighting.

90. ***The answer is A.*** Fire-resistive-rated glazing prevents the spread of flames and smoke, and also blocks the transfer of radiant and conductive heat. Beyond the requirements for needing to pass the hose stream test (for thermal shock) and size limitations as set by the manufacturer based on fire rating and testing, the glazing must be impact rated for use in hazardous locations (which a corridor would be) and must provide a barrier to both radiant and conductive heat transfer.

The certification of the manufacturer is not a factor, as it is the glazing itself that must pass the test required by the building code. The direction of the expected fire attack, while important for some products, is not a consideration because the glazing material should be able to provide protection regardless of the side on which a fire starts.

91. ***The answer is C.*** ASTM E119, *Standard Test Methods for Fire Tests of Building Construction and Materials*, is one of the most commonly used tests for the fire resistance of construction assemblies such as walls, floors, and ceilings. This method is also used for testing glazing to determine if the glazing meets a required hourly rating. For example, if a partition requires a 2-hour rating, any glazing in that partition must also be 2-hour rated.

16 CFR 1201, *Safety Standard for Architectural Glazing Materials*, is a standard for safety glazing used in hazardous locations. ASTM E84, *Standard Test Method for Surface Burning Characteristics of Building Materials*, is used to rate the surface burning characteristics of interior finishes and other building materials. It is also known as the Steiner tunnel test. NFPA 701, *Standard Methods of Fire Tests for Flame Propagation of Textiles and Films*, is used to test the flammability of draperies, curtains, or other window treatments.

92. ***The answer is B.*** A location plan, or key plan, is necessary to show how the proposed project relates to the exits of the building as a whole so the authority having jurisdiction (AHJ) can verify egress requirements, such as travel distance.

A furniture plan may be necessary for construction and installation but will not be required by the AHJ. A power plan would be developed by the electrical engineer and would be sufficient to show outlets and other electrical items. A sprinkler plan will be developed by the mechanical engineer or fire protection engineer for detailed construction. The interior designer may show sprinkler locations on the reflected ceiling plan to coordinate their locations with other ceiling-mounted items.

93. ***The answer is A.*** *The International Building Code* (IBC), in Chapter 9, prescribes minimum requirements for active fire protection, including detection, alerting occupants, controlling smoke, and controlling or extinguishing a fire, including requirements for portable fire extinguishers. While the same or similar provisions are contained in the *International Fire Code* and the National Fire Protection Association (NFPA) code and are referenced in the IBC, the IBC would be the easiest to use as most interior designers would have a copy of the code. The other codes are more complex and more difficult to use.

Specific requirements for buildings based on occupancy and use are found in Chapter 4 of the IBC. This includes voice/alarm systems required in some occupancies, such as high-rise buildings. Fire alarm and detection systems are found in Chapter 9 of the IBC, including where they are required in all the occupancy groups and where visible alarm notification appliances (flashing lights) are required.

Note that Chapter 7 of the IBC provides requirements for fire-resistive-rated construction, including how openings and penetration are protected.

94. ***The answer is D.*** ICC A117.1, *Accessible and Usable Buildings and Facilities*, is the latest edition of an ANSI standard that was first developed in 1961 and has served as the basis for subsequent rules, guidelines, and laws related to accessibility. The 2010 *ADA Standards for Accessible Design* replaced the original 1991 *ADA Standards*. 28 CFR 36, including its Appendix A, is the part of the *Code of Federal Regulations* that implemented Title III of the original Americans with Disabilities Act (ADA). The *ADA/ABA Guidelines* were the construction guidelines of the 2004 ADA law.

95. ***The answer is B.*** Occupant sensor controls are required in private offices, employee lunch and break rooms, conference rooms, copy rooms, and storage rooms, among other areas. An alternative is to use time-switch controls, but these would be awkward to use in an office application. Daylight responsive controls would not be effective if there was not enough area that could utilize daylighting. Toplight daylight controls would also not be effective if there is no toplighting.

96. ***The answer is A.*** Local zoning ordinances address the relationship of structures to their sites, and building codes address methods and materials of construction permitted within sites. A zoning ordinance defines what uses are allowed, where buildings may be constructed on the site, and how much building is permitted on a lot. Zoning ordinances can also address parking requirements, special requirements of historic districts or areas subject to a design or architectural review board, and site planning issues.

Information regarding maximum occupancy, fire-rated assemblies, and types of materials that may be used for construction is commonly found in building codes.

97. ***The answer is D.*** Although any of the options might be a possibility, the most restrictive requirement usually takes precedence when regulations are in conflict. Appealing a requirement takes time and costs money. Appealing a requirement is typically not worth the effort compared with designing to the most restrictive requirement. It could also delay the project.

98. ***The answer is A, C, D, and F.*** The use of materials with recycled content, those that can be reused, those with low embodied energy, and those that use renewable raw materials can all reduce the amount of resources in a building. Maintainability and low toxicity are important criteria for selecting materials but don't directly relate to material conservation.

99. ***The answer is D.*** The National Fire Protection Agency (NFPA) publishes NFPA 13, *Standard for the Installation of Sprinkler Systems,* and NFPA 13R, *Standard for the Installation of Sprinkler Systems in Low-Rise Residential Occupancies.* Both of these standards provide detailed requirements for the spacing, location around obstructions, and engineering for sprinklers.

The *International Building Code* (IBC) sets forth what occupancies and specific uses must have sprinklers but references the NFPA standards for installation details. The *International Plumbing Code* (IPC) does not regulate sprinkler systems, nor does the *International Mechanical Code* (IMC).

100. ***The answer is B.*** Occupant load determines means-of-egress requirements: how many exits must be provided to evacuate a space in case of an emergency, how large the exits must be, and where they must be located. Design occupant load for assembly occupancies is calculated under the *International Building Code* (IBC) by using the following requirements as given in Chapter 10 of the IBC.

standing space	5 net ft^2 per occupant
concentrated (chairs only)	7 net ft^2 per occupant
unconcentrated (tables and chairs)	15 net ft^2 per occupant

The largest occupant load determines the egress requirements.

101. ***The answer is C.*** The Business and Institutional Furniture Manufacturers Association (BIFMA) provides standards for many types of commercial office furniture, including office chairs, vertical files, desk products, panel systems, among others. The BIFMA standards define the specific tests to be used for each standard, the laboratory equipment that can be used, the conditions of the tests, and the recommended minimum acceptance levels. While each of the standard numbers is preceded by the American

National Standards Institute (ANSI) designation, the ANSI does not develop or write standards; they just approve standards development by other organizations, like BIFMA. An example is ANSI/BIFMA X5.1, *American National Standard for Office Furnishings—General Purpose Office Chairs—Tests.*

American Society for Testing and Materials (ASTM) does develop many standards for furniture flammability, safety, toxic substances control, durability, labeling, and similar standards, but it does not deal with the same issues as BIFMA. The National Fire Protection Association (NFPA) has some standards for furniture fabric flammability but does not deal with the construction and testing of individual types of furniture.

102. ***The answer is D.*** There are more than 49 occupants in this space, so two exits must be provided. These exits must be separate and a certain minimum distance apart so that a fire is unlikely to block both. To calculate the minimum size of the required exits from the space, multiply the maximum number of occupants by 0.2 in per occupant.

$$(300 \text{ occupants})\left(0.2 \frac{\text{in}}{\text{occupant}}\right) = 60 \text{ in}$$

For two exits, the minimum width per exit is 30 in. However, accessibility standards and component requirements both call for a minimum clear opening width of 32 in for each door. This width is generally achieved by using a 36 in door. Therefore, the best answer is two exits, two 36 in doors.

103. ***The answer is B.*** According to the *International Plumbing Code* (IPC), section 421.4, shower compartments cannot be less than 900 in^2 (0.58 m^2) in area and cannot be less than 30 in (762 mm) in least dimension. This means that the minimum size is 30 in by 30 in (762 mm by 762 mm). (One exception does allow a minimum dimension of 25 in (635 mm) provided that the minimum area is not less than 1300 in^2 (0.838 m^2).)

104. ***The answer is A, C, E, and F.*** The occupant load and the occupancy are the two primary determinants to establish whether two exits or exit access doorways are required. If the occupant load is above a certain number, then two exits or exit access doorways are required. More exits or exit access doorways are required when the occupant load is above 501. There are also limits on the common path of egress travel based again on occupancy and whether or not the building is sprinklered. Depending on how a space is planned, it may meet the requirements for needing only one exit; but if the common path of egress travel is exceeded, then two exits or exit access doorways may be required. All four of these variables are included in the *International Building Code* (IBC) Table 1006.2.1.

The common path of egress travel is the portion of the exit access travel distance measured from the most remote point of each room, area, or space within a story to the point where the occupants have separate and distinct access to two exits or exit access doorways.

The maximum travel distance is the distance that an occupant must travel from the most remote point in the occupied portion of the exit access to the entrance to the nearest exit. It doesn't affect the number of exits or exit access doorways required. The exit or exit access widths themselves depend just on the occupant load of the area served.

105. ***The answer is A.*** A local code official would be authorized to enforce all of the codes or standards listed except the Americans with Disabilities Act (ADA). The ADA is civil rights legislation that gives building users the right to sue the owner of a building if the users are denied access to a facility because the building is not accessible.

The other three options are model codes or standards that, when adopted by a local jurisdiction or state, become part of the building code. Depending on the jurisdiction, the code official could enforce ICC/ANSI A117.1, the *International Building Code*, or the *Life Safety Code* (which references ANSI A117.1).

❺ INTEGRATION WITH BUILDING SYSTEMS AND CONSTRUCTION

106. ***The answer is A.*** Option B is incorrect because a solid-core, wood door in a steel frame could be used to meet the conditions. Option C is incorrect because a smokeproof opening can be achieved with a wood door, as well as with a hollow metal door. Option D is a possible choice, but minimal maintenance under heavy use does not necessarily imply a metal door.

107. ***The answer is B, C, and D.*** Safety glazing is required in shower doors and glass panels less than 24 in (610 mm) from a door if the lower edge of the glass is less than 60 in (1525 mm) from the floor. Glass panels where the sill is more than 18 in (457 mm) above the floor do not have to be safety glazed. Glass panels with a conforming crash bar and glass panels less than 9 ft^2 (0.84 m^2) do not require safety glazing.

The illustration indicates, in summary form, where safety glazing is required.

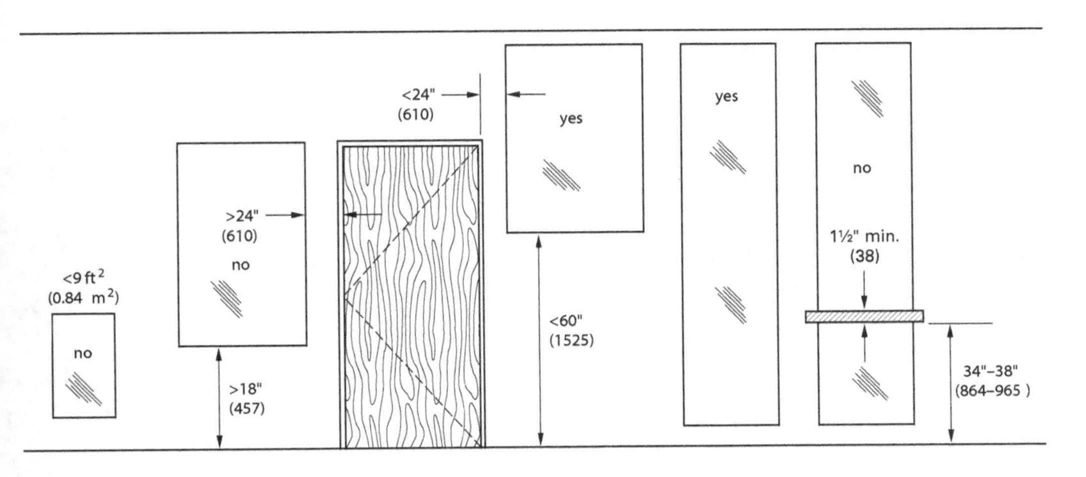

108. ***The answer is C.*** There should be at least 2 in (51 mm) of space between the inside face of the glass and the window covering to avoid excessive heat buildup, which could cause the glass to crack or break. There should also be sufficient space at the top and bottom of the window covering to provide for air circulation to prevent heat buildup. The mechanical engineer should review and approve any window covering detail to verify that there is not a problem with the location of the window covering or its interaction with the type of glazing that has been specified for the building.

Transparency refers only to the amount of light coming through the window covering and does not have as much effect on heat buildup as the space between the covering and the glass. The direction of airflow does not affect heat buildup, but it may affect the movement of the window covering when air is directed toward the window from a ceiling register. While the color of the window covering may affect heat buildup, it is not as critical a consideration as the spaces allowing for air circulation.

109. ***The answer is A.*** In any situation that involves or might involve a structural question, an engineer or architect should be consulted.

110. ***The answer is A.*** Card readers would be the most cost-effective solution because they could provide access control, as well as notification if there is an attempt at unauthorized entry.

111. ***The answer is C.*** Although all of the listed options *could* be modified to improve acoustic separation, a convector is the most troublesome and *must* be modified or it will allow sound to travel through the vents in one office, through the opening around the convector pipe, and out the vents in the adjacent office. Refer to the following illustration.

Illustration for Solution 111

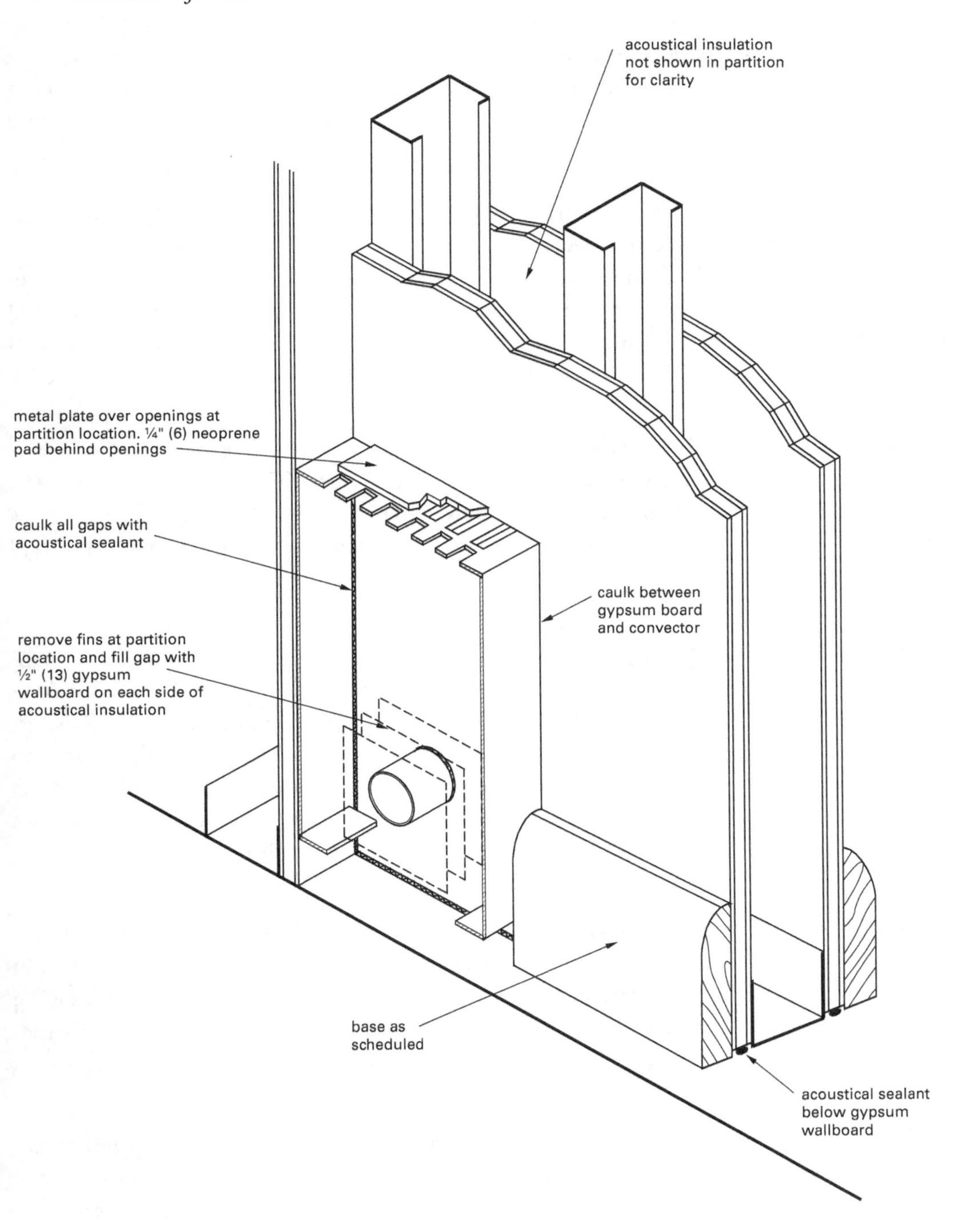

112. ***The answer is B, C, D, and F.*** Options B, C, D, and F are all important considerations in controlling both noise within a room and noise being transmitted from outside a room. A large ceiling might be useful for applying sound-absorbent material, but the size of the ceiling is already determined as required by the program. A white noise system is useful for masking sounds in an office environment, but would not be appropriate in a classroom due to its cost and limited benefit in masking the types of noise in a classroom.

113. ***The answer is A.*** An arc-fault interrupter detects arcing of electricity, such as that caused by faulty wiring of a lamp, and de-energizes the circuit. Arc-fault interrupters are designed to address fire hazards. Both arc-fault interrupters and ground fault interrupters can be installed as a receptacle or as part of a circuit breaker. However, arc-fault interrupters are also required to serve lighting, so they are generally placed in the breaker box.

A circuit breaker detects overcurrents and operates much more slowly than an arc-fault interrupter. A disconnect switch must be operated manually. Ground fault interrupters (also called ground fault circuit interrupters) detect small leaks of current to an unintended grounding source, such as when someone touches a faulty device and a water pipe, and de-energize the circuit. They are designed to address shock hazards and are required in areas near water, such as bathrooms, kitchens, and laundry rooms.

114. ***The answer is D.*** Neither option B nor C is correct, because wire and float glass, which do not meet the requirements of either ANSI Z97.1 or 16 CFR Part 1201, are not considered safety glazing. Option A is also incorrect. Although there are some types of laminated ceramic glazing that meet the requirements, most are not considered safety glazing.

115. ***The answer is D.*** A post-tensioned concrete slab contains tendons under high tension. Normally, these tendons are located throughout the slab, which would make it very difficult, if not impossible, to drill a small hole through the structure. Concrete over steel decking has little or no reinforcement and is easy to core drill. A flat slab concrete slab has reinforcing, but that reinforcement can be located with X-raying, and cutting one reinforcing bar is not usually a problem. With a one-way pan joist system, it is easy to locate the joists and drill through the thinner slab.

116. ***The answer is B, D, and F.*** The interior designer can design partition slip joints for floor deflection, ceiling systems for seismic events, and mullion slip joints for wind sway. The architect or structural engineer must design movement joints for building expansion into the original building structure. Movement caused by foundation settlement and thermal expansion cannot be accommodated by interior design details.

117. ***The answer is A.*** The opening can be framed, but a structural engineer will be needed. Option B is incorrect because an architect would have to do more than just sign the drawings; structural calculations would be necessary. Option C is incorrect because the interior designer is not qualified to determine what type of header is needed. Option D is incorrect because the opening can be made as long as it is engineered properly by a qualified professional.

118. ***The answer is A.*** A beam and girder steel system would most likely be covered with a concrete slab on steel decking. This would be the easiest to penetrate for a large opening between the beams. Even if one beam had to be removed to provide a larger opening, additional reinforcing beams could be installed to frame the new opening.

Flat plate concrete contains a dense grid of reinforcing steel. Cutting the existing steel would weaken the overall structure too much (although it would be technically possible to reinforce the opening). A hollow-core concrete slab is also integrally reinforced, and cutting a large opening would weaken the overall span. Open web steel joists span longer distances, with spacing from 2 ft (610 mm) to 6 ft (1829 mm). This would limit the size of the opening to the joist spacing and would probably not be enough for a stairway.

119. ***The answer is A.*** An all-air system can be subdivided into as many individually controlled areas as needed. Radiant panels could be used, but it would be awkward to locate them in the ceiling and the cost would be much higher than that of an all-air system. Radiant panels also would not provide for cooling.

120. ***The answer is A.*** Upright sprinklers disperse the water upward so coverage is provided above and below the suspended wood-slat ceiling.

121. ***The answer is C.*** Drain lines need a minimum slope to function properly. For example, a 3 in (76 mm) drain line from a toilet would need a minimum slope of ⅛ in (3 mm) per ft (300 mm). The distance that the new toilet would be located from an existing drain would determine the total vertical distance needed to allow the drains to function correctly.

Water supply lines can be any length. A single cold water line could be used with an instantaneous water heater at the sink, if necessary. Floor-to-floor height is not a limiting factor. The dimensions of the plenum space could limit the location if the drain needed to be incorporated totally within the plenum, but the ceiling below could be lowered slightly or a drain pipe could be furred in if necessary.

122. ***The answer is C.*** The interior designer uses information about the client's needs and requirements of the furniture systems to determine where telephone and data connections should be made. The building architect (option A) is not involved with these decisions. The electrical engineer (option B) and the telecommunications consultant (option D) are given the locations of various types of outlets by the interior designer, and then perform the necessary selection of wiring and sizing of conduit needed for connections to central stations.

123. ***The answer is D.*** A minimum slope is required for toilet drainage, which limits how far a toilet room may be from the nearest soil stack.

Cold water is supplied through underground pipes, and their location is not a limiting factor. Only an air connection is required to the vent stack, so the location of the vent stack is not a factor. A toilet's location could be affected by the location of a structural beam below it, but not by one above it.

124. ***The answer is C.*** Extending the partition through the ceiling is the best method because it provides the most mass and fewest joints. Ceiling tiles with high CAC ratings do not provide the same sound attenuation as a partition. Using sheet lead may help a little, but sealing the joint between the suspended lead sheet and the top of the ceiling assembly is difficult and not always effective. Suspending framing with one layer of gypsum board also makes it difficult to seal the joint between the suspended elements and the top of the ceiling tiles. Further, one layer of gypsum board is not as effective as two layers.

125. ***The answer is A.*** Flat plate construction consists of a single thickness of reinforced concrete supported by concrete columns. Except for the columns, there are no structural elements below the underside of the slab, making it possible to place needed services as high as possible, thereby allowing for the highest finished ceiling.

A flat slab includes concrete drop panels (either square, round, or cone shaped) around each of the columns to increase strength. These drop panels prevent the installation of services near the columns, decreasing the possible finished ceiling height. Both open web steel joist and waffle slab systems occupy much more space than a flat plate slab and would require a lower finished ceiling height after services are installed.

126. ***The answer is B.*** When an astragal is used to seal the opening between double doors and the doors are fire rated, a device known as a door coordinator is needed to automatically close one leaf of the pair of doors before the leaf with the attached astragal closes. This is to prevent incomplete closing of the doors.

An automatic door bottom and threshold are not necessary. Panic hardware is only required in specific situations defined in the building code, where the double doors are exit doors in occupancies with large numbers of people.

127. ***The answer is D.*** Because of its long, thin shape and its adjustability, a slot air diffuser would provide the best air flow along a window wall. A round diffuser would distribute air in all directions, including away from the wall. Square diffusers provide distribution from only one isolated location.

128. ***The answer is C.*** A partition that penetrates a suspended ceiling by just a few inches is typically used when the ceiling heights on both sides are different, when the ceiling grid changes direction, or when the ceilings are different materials. In each case, the gypsum wallboard extending above the ceiling provides a base for the attachment of the ceiling angle. Either the studs can be extended to the structure without wallboard, or the top of the partition can be framed with a runner, which is then braced with diagonal studs anchored to the structure above.

The partition would not provide any additional sound control because of the lack of wallboard in the plenum. Shelving could still be installed if the partition extended only to the suspended ceiling and were attached to the ceiling grid; the framing does not need to continue to the structure. The detail shown does not provide the required suspended ceiling configuration for seismic restraint.

129. ***The answer is B.*** Although all the information listed in the options needs to be included, the most important is the clearance provided near the ceiling to allow the paneling to be installed. The installation of the panel is not really affected by the thickness of the wood cleat or by the size of the base.

130. ***The answer is A.*** The detail shown suggests a custom-made unit. This drawing indicates a custom-fabricated piece of architectural woodwork. Finish carpentry is completed on the jobsite and seldom includes complex items such as bookshelves, desks, and paneling. Manufactured casework is prefabricated cabinetry built off site to standard dimensions, such as residential and institutional cabinets. Standard cabinetry is not a term used in interior design and architecture.

131. ***The answer is C.*** The majority of a cabinet is built with $^3/_4$ in (19 mm) panels including the bottom, sides, and top bracing. The back is typically $^1/_4$ in (6 mm), but this was not one of the options.

❻ INTEGRATION OF FURNITURE, FIXTURES, AND EQUIPMENT

132. ***The answer is B.*** Because the furniture dealer and interior designer are the two people closest to the specification and ordering of furniture, option B is correct. Furniture manufacturers, general contractors, and clients are seldom, if ever, involved in budgeting furniture.

133. ***The answer is B.*** A drop ship order is simply a purchase order requesting that merchandise be delivered somewhere other than the address of the person or company ordering the merchandise.

134. ***The answer is A.*** The first choice is the simplest and most reliable because it puts the entire burden of matching the existing finish on the contractor and painting subcontractor. They are the people most likely to have the knowledge and experience to make the match. Also, putting the notes on the drawings and specifications ensures that the general contractor would be responsible for correcting the finish if it did not match.

135. ***The answer is C.*** Metal-halide lamps are the best choice because they have a high efficacy (80 lm/W to 120 lm/W) and good color rendition. These two advantages would make metal-halide a better choice, even though mercury-vapor lamps have a longer life.

136. ***The answer is B.*** Because furniture dealers actually sell the furniture to designers at whatever trade discount is available, they would be the best source for current, accurate prices. Furniture manufacturers sell only to the dealers and only suggest retail prices. The internet would not have the prices, including trade discounts, that are available to the designer. A merchandise mart leases space to dealers and is not involved in pricing for the merchandise.

137. ***The answer is C.*** Systems furniture that uses panels of various lengths that link together and provide support for work surfaces and storages are the most effective to provide flexibility for open office planning. Ready-made furniture consists of fixed elements that may not be appropriate for changing needs. Moveable panels offer some flexibility, but with ancillary pieces, it is limited to what is originally specified. Custom design furniture is too expensive and not flexible enough for this use.

138. ***The answer is A, B, C, and F.*** Fabric needs to be permeable to allow sound waves to pass through. Hydrophobic fabric does not absorb and hold moisture to avoid sagging and distortion. Loose core material such as fiberglass or polyester batting should be used for sound absorption. Finally, the core material should be unbacked to allow sound waves to pass through. Tackable material, such as mineral fiberboard, can be used for tackability, but this was not part of the question. Fabric color is not necessary when designing the acoustic panels for sound control.

139. ***The answer is B, C, D, and E.*** Demountable partitions systems offer great flexibility and reuse potential. Because of this, they have a low life-cycle cost compared with built-in-place wallboard partitions and offer related components, such as door frames, glazing frames, bank rails, and other openings.

The demountable partitions are assembled in pieces, so there are small vertical joints between the panels. They do have a higher initial cost than standard partitions.

140. ***The answer is A, B, D, and F.*** Keynote numbers are required to identify the items on the furniture plan for reference to the schedule. A description is needed as a general name for the item.

Both the manufacturer name and manufacturer's catalog number are required to specify the item. Size and color are not necessary, as they are defined by the item number.

141. ***The answer is D.*** The Wyzenbeek test determines the abrasion resistance of woven fabrics. A sample is mounted on a testing machine and abraded at a rate of 5000 double rubs per hour. Fabrics are rated on the number of double rubs they can resist and are classified as either light duty, medium duty, or heavy duty.

The Martindale abrasion test is used for the resistance of textile fabrics, generally with a pile depth of less than 0.08 in (2 mm). The Taber abraser test is commonly used for carpet. The tearing strength test measures the tearing resistance of fabrics after an initial cut has been made in the fabric. While this may be a test that could be included in the furniture specifications, it is not as important as testing for wearability.

142. ***The answer is B.*** Ergonomics is the most important consideration because it involves the application of knowledge of human physiology to the design of the physical environment. Fixture display design would require knowledge of factors such as human size, reach, and vision, and how they would affect the details of the fixtures.

Anthropometrics is the measurement of the human body and simply forms the basis of the ergonomic design of the fixtures. Gestalt is related to theories of perception, such as grouping and closure, and would not be a critical factor in designing the fixtures. Proxemics relates to how people use space, and may suggest what the size and spacing of the fixtures should be.

143. ***The answer is A, B, C, and E.*** A clear floor space of at least 36 in (915 mm) is required between tables and other displays. In addition, any tables would have to meet accessibility requirements for the height of the table. Knee space would be required to accommodate wheelchairs, at least around the customer portion of the table. Finally, reach ranges would have to meet accessibility ranges. Equivalent facilitation would only be required for any displays or shelving that were out of reach. Tolerances would not be a critical factor because the tables would be fabricated in a woodworking shop and would be precisely made according to the designer's drawings.

144. ***The answer is C.*** The interior designer completes a purchase order (or more than one if required) listing the items to be purchased, their catalog numbers, prices, color, and other data. The purchase order is sent to the manufacturer or vendor of the furniture. A sales agreement, which is basically a contract signed by the client, obligates the client to pay for the items listed in the agreement. An acknowledgment comes from the manufacturer to the designer, repeating the information on the purchase order along with costs, shipping date, and how the consignment will be shipped. The freight billing relates to transportation and is handled by the manufacturer, vendor, dealership, or others.

145. ***The answer is D.*** The riskiest method for the designer for procuring furniture is to act as a reseller, much like a furniture store. This also requires a resale license. In this scenario, the designer helps the client select the furniture, writes purchase orders, accepts delivery, arranges for installation, and collects money from the client (including taxes), which is then used to pay the manufacturers or vendors. The designer makes money with this method by buying the furniture at a significant discount (wholesale) and then charges the client a higher amount, often retail price. The designer runs a risk throughout the entire process as many things can go wrong, including nonpayment by the client.

As a purchasing agent, the designer writes the purchase orders and completes other paperwork but the client is responsible for making payments. Writing specifications and turning the entire process over to a dealership is the simplest method of procurement and the least risky. Ordering the furniture and hiring an expediter is not a typically used method.

146. ***The answer is A.*** Items in the furniture, fixtures, and equipment (FF&E) budget are specified, purchased, and installed differently than construction items. The FF&E items may have different discounts than the designer or client receives for the construction items and may require different taxes depending on where they are sourced. In the case of construction (partitions, doors, hardware, etc.), the designer specifies individual products, which are estimated by the contractor, priced out along with overhead and profit, and the amounts billed to the client. Sub-contractors and sub-subcontractors may also be involved with their own overhead and profit.

147. ***The answer is B.*** Attic stock is a term used to describe extra materials that are required to be purchased under a project's original contract and that are stored for later use. The purpose is to keep a supply of material that can be used as replacements for worn or damaged products during the life of the project, which will match the style, color, and finish of the original product. They are kept in on- or off-site storage so that a substitution can be made quickly without having to order from the supplier. Examples of commonly specified attic stock include flooring materials, ceiling tiles, door hardware, upholstery fabric, and system furniture components. Attic stock may also even include furniture items, such as in hotels, where a furniture item may need to be replaced quickly to avoid shutting down an income-producing guest room.

Any leftover materials are the responsibility of the contractor. Perishable materials should not be ordered as attic stock. Paint, for example, can be ordered from the factory to exactly match the original paint. High-cost furniture would probably not be purchased due to the extra cost in both purchasing and storage involved unless any facility expansion would be imminent, or in the case of hotels as mentioned above, would be needed to maintain income-producing areas.

148. ***The answer is A, B, D, and F.*** In the life-cycle cost analysis method, maintenance costs, operational costs, and taxes should be included with those costs set at the present value of the base date. Residual value at the end of the study period should be subtracted from the other costs. Options C and E are incorrect because the cost of any replacements and finance costs should be discounted back to the base date to account for the time value of money.

149. ***The answer is A, B, E, and F.*** Chandeliers are fixed lighting fixtures and would be part of the electrical contract budgeted and paid as part of the construction budget and installed by the electrical contractor. Wood cabinets are architectural woodwork and are part of the construction budget. The other items are considered FF&E.

❼ CONTRACT ADMINISTRATION

150. ***The answer is A.*** The certificate of occupancy (CO) (also called the letter of occupancy or use and occupancy letter in some jurisdictions) is issued by the authority having jurisdiction, which allows the client to occupy the space.

151. ***The answer is A.*** A manufacturer issues an acknowledgment after receiving a purchase order. Because the information on the purchase order is repeated, the designer can compare this with the original purchase order to make sure they match.

152. ***The answer is D.*** The contractor should make sure that the necessary samples, shop drawings, and other required submittals are forwarded to the interior designer for review. Because the general contractor is responsible for coordinating the various trades and suppliers, he or she would be responsible for the mechanical shop drawings.

153. ***The answer is C.*** The interior designer's review of the shop drawings is only for conformance to the general design intent of the job. The general contractor is responsible for coordinating the job, checking dimensions, and in general, building the job according to the contract documents.

154. ***The answer is B.*** The interior designer is responsible for designing the job according to governing building codes. The contractor often points out problem areas ahead of time, but he or she is under no obligation to do so.

155. ***The answer is A.*** That a contractor should notify the designer of the discrepancy in writing is standard procedure that is written into most general conditions, including the AIA Document A201, *General Conditions of the Contract for Construction.*That a contractor should notify the designer of the problem and suggest a remedy is close to being correct, but the response does not specify whether or not the contractor notifies the designer in writing. In addition, although the contractor often suggests how to solve a problem, that is the designer's responsibility.

156. ***The answer is B, C, D, and E.*** An addendum is issued only before the construction contract is signed, it is used to modify the contract documents, it must be filed with the other bidding documents, and it must be issued at least four days before bid opening. Addenda must be sent to all the contractors bidding on a project. Addenda are only issued before the contract is signed.

157. ***The answer is A, B, and D.*** Although most bid openings are open to anyone who wants to attend, the owner, the interior designer, and the general contractors bidding on the project are the parties most commonly in attendance. Material suppliers and lending agents generally do not attend the bid opening. On particularly large projects, major subcontractors, such as mechanical and electrical, may also attend.

158. ***The answer is C.*** The contractor should always request approval in writing so the interior designer can review the information about the tile and make a determination whether the specification was an "or equal." If the request is approved, the interior designer will issue an addendum to all contractors telling them that the new product was approved.

159. ***The answer is C, D, E, and F.*** Verifying contractual obligations, reviewing consultant billings, writing proposals, and coordinating consultants are all part of a project manager's duties.

It is not the responsibility of a project manager to hire new staff to work on a project. Although a project manager may suggest that new staff be added to a project and may sit in on interviews when appropriate, the actual hiring is the responsibility of the firm principal or the human resources director. Likewise, the project manager is not responsible for reviewing the office-wide aged accounts receivables; this is the job of the firm principals or the office accountant (although the project manager is concerned with the payment history of the jobs).

160. ***The answer is D.*** During the planning stage of a project, a project manager is responsible for setting requirements in three areas: time, fees, and quality. With time planning, the project manager schedules the work required and ensures that there are enough funds and staff to complete the project. Fee projection looks at the project's scope, likely costs (e.g., direct and indirect expenses, overhead, consultant fees, and reimbursables), and the desired profit in order to allocate enough money to complete the project within the allotted time. Quality planning involves determining client expectations. Monitoring, directing, scheduling, and coordinating are all tasks done after the planning stage and throughout the project's progress.

161. ***The answer is D.*** All project-related work and communication should be documented, filed, and saved. A project notebook contains only those documents needed on a daily basis. In addition to standard forms and written correspondence, items such as meeting notes, telephone conversations, on-site decisions by the client, and programming information should be included in the project documentation. Many documents that have no legal significance can still be important for recording decisions and other project facts, or may be useful to the design firm on future projects.

162. ***The answer is A, B, D, and E.*** The project manager organizes the design schedule based on client requirements and office staffing, and sets the time allowed for bidding and when bids are due. If the project requires specifying furniture, the project manager is also responsible for scheduling the furniture's order and delivery. If predesign programming is part of the design work, the project manager would also schedule that. Only the contractor is responsible for the construction and commissioning schedule.

163. ***The answer is A.*** A Gantt chart is another name for a bar chart.

164. ***The answer is A, C, D, and E.*** The client's approval process, the design firm's staff size, agency approval requirements and schedule, and the project complexity can all affect the overall schedule, especially on large projects where many levels of client reviews and approvals may be necessary.

Even if the client has a dedicated facility management department, the size of the client's staff has little effect on the speed with which the interior designer's office can complete its work. The design methodology of senior staff will not have an influence on project completion time, as this is already taken into account in the overall project schedule.

165. ***The answer is B, D, and E.*** Follow-up makes a good impression on a client, can help solve minor problems after move-in, and can be used to gather information for future projects.

Additional follow-up after a project's completion is not contractually required by either the standard American Institute of Architects (AIA) documents or the American Society of Interior Designers (ASID) forms for interior design services. Although professional photographs of the project are commonly arranged by the interior designer, they are not required. Post-occupancy evaluation is not required under standard contracts; it is an additional service.

166. ***The answer is A, D, E, and F.*** Keeping notes, planning and assigning tasks for staff, monitoring fee expenditures, and keeping up with the project's progress are all common responsibilities of the project manager. Although on small projects, the project manager may organize drawing layout, this is usually the task of the job captain or the person in charge of preparing the drawings. Long-term goals are the responsibility of the firm's principals.

167. ***The answer is B.*** The interior designer is responsible for knowing the size of built-in items and for designing and detailing construction into which those items are placed.

168. ***The answer is C.*** Furnishings are bought through the purchase order process. Releasing funds for construction is done with an application and certificate for payment.

169. ***The answer is C.*** Retainage is a percentage of each payment that is withheld by the owner and not paid until the contractor has completed their work.

170. ***The answer is A, B, E, and F.*** The appropriate members of the team should attend all project meetings that should also be recorded for the complete topics discussed. The recordings should be distributed to all those attending and to the client. As part of the distribution of notes, there should be a statement that gives the participants the opportunity to add or correct the notes. Notes can be either handwritten or keyboarded directly by a designated person attending the meeting and later edited for clarity. There is no need for a statement that the notes represent the final results nor is there is a need for them to be handwritten.

171. ***The answer is D.*** The owner is ultimately responsible for paying for any mock-up, as they are part of the specifications. If a mock-up is acceptable, it is incorporated into the construction. If not, it must either be corrected, if possible, or discarded.

172. ***The answer is C.*** The interior designer should return the samples to the subcontractor without reviewing them. The subcontractor should first submit them to the contractor for review for compliance with the construction documents, in accordance with AIA Document A201, *General Conditions of the Contract for Construction*, if these documents are used. Only after this review should the contractor send them to the interior designer for review. If they are not checked and signed by the contractor, the interior designer should immediately return them without review.

173. ***The answer is C.*** The cost of labor and materials is $22,225. The addition for overhead and profit is 20%.

$$(\$22,225)(1.20) \quad = \$26,670$$

The addition for coordination is 5%, and this is applied after the addition for overhead and profit, for a total of,

$$(\$26,670)(1.05) \quad = \$28,004 \quad (\$28,000)$$

174. ***The answer is B.*** An addendum is used to make changes to the contract documents after they are issued for bidding but before the contract is awarded. Change orders and construction change directives modify the original contract documents after the contract is awarded. An alternate listing is simply the list of alternates that the contractor must include in the bid.

175. ***The answer is A.*** A mock-up is a full-sized sample of a portion of the construction, commonly built on the jobsite. It can be either separate from the building or, if approved, can be integrated into the building. A mock-up is called for in individual sections of the specifications for items that will be repeated throughout the project, innovative construction techniques, and other instances where it is useful to see the "finished product" before it is truly finished. Mock-ups can be very expensive, so they should only be specified where necessary. They become more cost effective if the approved mock-up can then be integrated into the building. An example would be a mock-up for a complex piece of storage millwork that would be repeated many times on a large project.

Product data, samples, and shop drawings may all be provided by the contractor for the interior designer's review in the designer's office.